SPACES OF SOCIAL EXCLUSION

In all developed countries, to varying extents, a minority of the population is widely perceived as suffering from particular deprivation and disadvantage. The European Union and many national governments and local agencies have sought to conceptualise and deal with this problem through the notion of 'social exclusion'.

Spaces of Social Exclusion explores the forms of social exclusion in the developed countries, its social and spatial causes, and the role of space in political strategies to address it.

Part One introduces historical and contemporary ideologies of poverty and social exclusion and strategies towards them. It describes the social and spatial patterns of disadvantage in advanced capitalist countries. Part Two analyses the origins of social exclusion by examining the different spheres of disadvantage and their relations, emphasising the role of space, place and scale. It brings together and integrates research on diverse aspects of social exclusion and the varied processes which produce it. Part Three presents and criticises strategies for overcoming social exclusion from across the political spectrum, focusing particularly on their uses of space.

This accessible but theoretically rigorous book demonstrates the similarity throughout the advanced capitalist countries of the processes which create social exclusion. It shows how space is used in reproducing exclusion but can also be used to contest it.

Jamie Gough is Senior Lecturer in Town and Regional Planning at Sheffield University.

Aram Eisenschitz is Senior Lecturer in the School of Health and Social Science at Middlesex University.

Andrew McCulloch is Senior Lecturer at Northumbria University.

SPACES OF SOCIAL EXCLUSION

Jamie Gough and Aram Eisenschitz with Andrew McCulloch

Routledge
Taylor & Francis Group

LONDON AND NEW YORK

First published 2006
by Routledge
2 Park Square, Milton Park, Abingdon, Oxon OX14 4RN

Simultaneously published in the USA and Canada
by Routledge
270 Madison Avenue, New York, NY 10016

Routledge is an imprint of the Taylor & Francis Group

Typeset in Garamond by
Keystroke, Jacaranda Lodge, Wolverhampton
Printed and bound in Great Britain by
TJ International Ltd, Padstow, Cornwall

British Library Cataloguing in Publication Data
A catalogue record for this book is available from the British Library

Library of Congress Cataloging in Publication Data
A catalog record for this book has been requested

ISBN10: 0–415–28088–5 (hbk)
ISBN10: 0–415–28089–3 (pbk)

ISBN13: 9–78–0–415–28088–4 (hbk)
ISBN13: 9–78–0–415–28089–1 (pbk)

CONTENTS

ILLUSTRATIONS

FIGURES

TABLE

PREFACE

There are many good books on poverty in the developed countries, and now quite a few on social exclusion. Why, then, write another text? In this book we have aimed to develop an approach which is novel in three respects.

- Existing books on poverty are mostly within the tradition of social policy. They describe poverty in a given society at a particular time, and examine state policies explicitly aimed at alleviating it. In this book we have widened the focus to examine the creation of poverty and social exclusion by economic, social, cultural and political processes in space. Our analysis links to debates about the nature of the present epoch and its spatial forms – post-industrialism, the information society, the networked society, post-Fordism, postmodernism, neoliberalism, globalisation. This deeper analysis of the causes of exclusion leads us to examine a wider range of strategies against poverty and to scrutinise their theoretical assumptions.
- Texts on poverty in the developed world tend to assume national societies and national states as the primary objects of study; variations in poverty within the national space, and local policies towards poverty, are also sometimes considered. Our view is that geography is both more important and more problematic than this; we have therefore given space more space. As well as investigating spatial variation in exclusion within and between countries, we draw out the active role of space, place, scale and distance in creating exclusion, and the intrinsically spatial nature of the state's interventions. We argue that scales from the home to the globe and, crucially, their articulation are germane to both the creation and overcoming of poverty.
- General texts on poverty usually consider a certain range of policy options from minimal relief of poverty to stronger and more egalitarian interventions. This book is more politically inclusive. We examine the implications for poverty and exclusion of strategies from the far right to the far left. We take seriously the appeal of each strategy, tracing their social and spatial logics and their historical emergence, and draw out their fundamental tensions. Our aim is to sharpen the political debate and to show the importance of space within it.

The authors would like to thank Kim Smith for her help with sources, John Veit Wilson for discussions of social policy, and Paul Langley for detailed comments on a draft of Chapter 5; none of these, of course, necessarily agrees with what we have written here. Our thanks also to Yvette Brown for help in finding photographs; Duncan Fuller for permission to use a photograph of his; and Sunny Cheung for dropping everything to help prepare the pictures. We are very grateful to our editors at Routledge, Andrew Mould and Zoe Kruze, for their patience and generosity in dealing with our delays and mistakes and in accommodating our quirks. Finally a cliché, but a truth: we could not have produced this book without the support of John, Lorraine and Cecelia, and we are enormously grateful to them.

ABBREVIATIONS

Asbo	Anti-Social Behaviour Order
BME	black and minority ethnic
CBD	central business district [of a city]
CI	conservative interventionist/m
EZ	Enterprise Zone
ILO	International Labour Organization
LETS	Local Exchange and Trading Schemes
LSP	Local Strategic Partnership
MDC	more-developed country
PB	participatory budget
RDA	Regional Development Agency
SERPS	State Earnings-Related Pension Scheme
SRB	Single Regeneration Budget
UDC	Urban Development Corporation
WTO	World Trade Organization

INTRODUCTION

THE COMING OF 'SOCIAL INCLUSION'

My vision is of a nation where no-one is seriously disadvantaged by where they live, where power, wealth and opportunity are in the hands of the many not the few. [Our aim is] one nation, not separated by class, race, or where people live.

[handwritten margin note: ? Separated by what then?]

(Tony Blair in Social Exclusion Unit, 2001a)

The Labour government which came to power in Britain in 1997 after 18 years of Conservative rule declared itself committed to fighting an evil with a new name, 'social exclusion'. Whereas previous Labour governments might have railed against the *poverty* created by their Conservative predecessors, the Blair government framed the issue in a new way. To be sure, the government sometimes refers to poverty in the traditional sense, that is, a low income compared to the national average, and has introduced a number of measures to increase the income to certain groups. But, borrowing a term first used by the EU, its preferred concept is social exclusion. This concept frames the problem in a particular way, as one of *integration* of a *distinct* population into mainstream society. There is a new underclass which has become detached from normal social life. It is this detachment which accounts for their poverty; these problems cannot be overcome without a transformation in the *form of life* of the socially excluded. The political task, then, is to draw this underclass into the mainstream. The central element of this normalisation is waged employment; but it also involves the excluded changing their family life, public behaviours, community participation, attitudes to the self, and involvement in politics – in short, a cultural as well as an economic transformation.

The conceptual shift from poverty to social exclusion not only offers an explanation of poverty but makes the problem itself more complex. What is now undesirable is not (just) lack of money but, potentially, a whole set of qualitative problems: poor housing, ill health, educational failure, commission of crime or being its victim, use of narcotics, failure to participate in conventional politics, and so on. These are not simply products of poverty but social problems in their

1

own right. Moreover, people in different social groups and situations may experience these problems quite differently, which may thus need addressing in distinct ways. Redistribution of income to 'the poor' is therefore not enough, or even necessarily desirable. Indeed, low income as such may not be a problem providing the person is well integrated into social networks. The poor pensioner with a good network of friends and the unemployed person with a strong ethnic-community network are not socially excluded and may therefore not be disadvantaged or a social problem. The key issue then becomes social isolation or social deviancy *rather than* low income.

This is a novel ideology. Very different premises underlay the social democratic approach towards poverty which developed from the late nineteenth century, took full form in the Beveridge settlement after the Second World War, and expanded up until the 1970s. Poverty was regarded as a misfortune occasioned by the vagaries of a market economy. A safety net of basic benefits should be provided by the state, though individuals were to supplement this from their own resources or through their family. After the Second World War, the national state offered to guarantee full employment. Good quality housing should be available to the poor through state provision, and the basic services of education, health, social care and leisure amenities would be provided at a *universal, national* standard and for the most part free at the point of use. The state ran the utilities and public transport partly in order to provide a standard, affordable service. The national universalism of this strategy was frequently betrayed by the influence of markets and distortions by social power; nevertheless, it did inform the structures of provision and, just as importantly, was a promise about how things *should* be.

Both ideology and practice changed dramatically in the 1970s and 1980s with the rise to dominance of a new social settlement, neoliberalism. Neoliberalism portrayed the previous welfare regime as expensive and wasteful, as a disincentive to provide for one's self, as nationally-centralist and dictatorial, and as sapping individuals' enterprise and independence and limiting their choice. The state's role in income support and public services was cut back. Blockages to markets such as trade unions, excessive taxes and subsidies were attacked. The key scale of social regulation is thus shifted from the nation to the individual operating in markets of varied scale. Poverty was to be overcome principally through economic growth, generated by better resource allocation through market mechanisms and by unleashing enterprise. Free labour markets would ensure that the working-age poor would find employment provided they accepted the market's verdict of their worth. Poverty would wither away.

In the event, the reverse occurred. Poverty increased massively, especially of people permanently on (deteriorating) state benefits and pensions, the (enormously increased) number of unemployed, and those in (increasingly) low-paid and insecure jobs. A large number of the unemployed were effectively cut off from the labour market by social and spatial barriers. The most visible, and therefore troublesome, section of the poor were those in the sink estates and deteriorating housing of the cities and large towns where a high proportion of manual jobs had disappeared. The urban poor were increasingly seen by middle-class opinion as being an unruly and potentially dangerous 'underclass', useless for the economy and a drain on it. This view of the poor is nicely expressed by Tony Blair in the text quoted above:

below

> [W]e inherited a country where hundreds of neighbourhoods were scarred by unemployment, educational failure and crime. They had become progressively more cut off from the prosperity and opportunities that most of us take for granted. Communities were breaking down. Public services were failing. People had started to lose hope.

This was the starting point for the notion of 'social exclusion'. A free market economy is indeed the most dynamic and the best way to raise average incomes that 'most of us take for granted'. But this very dynamism means that some people inevitably get left behind. This is not entirely their fault. Complex combinations of household form, health, failings in education, poor environments and labour market shocks can send individuals and groups into downward spirals of decline. These are compounded by concentration of the poor in particular neighbourhoods causing 'communities to break down'. There is therefore a need for intervention to reverse these vicious circles of decline; accordingly we term this strategy 'conservative interventionism' (CI). Intervention may be needed in any of the numerous fields in which disadvantage is experienced. But it is aimed above all at integrating those of working age into waged employment, seen as both the means and the key sign of social inclusion; this realises the neoliberal aim of self-reliance. Since neighbourhoods compound and indeed foster exclusion, intervention must be targeted on them. Since the state is heavy-handed and inflexible, intervention is to be via decentralised local partnerships between parts of the state, quangos, residents and business. These steer the poor away from 'the dependency mentality' towards active participation and taking responsibility for their own welfare. The social inclusion strategy thus accepts the main thrust of neoliberalism but seeks to make it function better through soft, spatially dispersed and differentiated forms of intervention. Like neoliberalism it rejects the 'monolithic' universalist welfare state, but seeks to amend some manifest failings of the neoliberal project.

This development of this strategy since the early 1990s and its current dominance were the immediate stimulus for this book, and form a major focus of it. We seek also to understand and critique the social democratic settlement and its neoliberal nemesis which led into social inclusion. There are alternatives to the left of these strategies which we regard as more promising for the poor and excluded, and which have had a significant impact historically and in the present. Indeed, the dominant strategies of the last half century can be viewed as means of heading-off these left alternatives. We consider two of the latter in particular, the associationalist strategy, currently the most influential re-working of social democracy, and a socialist strategy of self-organised opposition to capital and social oppressions. We shall argue that particular uses of space are integral to all these strategies. We thus seek to politicise – and spatialise – more fully the debate on exclusion.

SLIPPERY TERMS, REAL LIVES

Political strategies have changed the concepts used to denote disadvantage. The critical observer has to accept these terms in order to engage with these strategies, but also to have some distance from them in order to attempt to get closer to people's lived experience. What words, then, are we to use in this book? 'Social exclusion' is a potentially fruitful term in a number of ways. First, it suggests that *social relations* other than income may exclude, for example gender or racism. Second, social exclusion points to the way in which the poor and disadvantaged are excluded from important types of *social interaction and social activity* (Sen, 1983). Understood in this way it takes further the notion of 'relative poverty' – that the adequacy of income should be judged according to the norms of life of the given *society* (Townsend, 1979).

Social exclusion can, however, be used misleadingly. First, it can play down the importance of *income and material resources*. As a result, the problematic of *inequality* is occluded, since this is most clearly measured by income or wealth. The poor may then be 'excluded', but not by the wealth or business of the rich. Second, the term is ambiguous on the political relation between the individual and society. 'Social' may suggest that disadvantage is *socially created*, directing us to an analysis of the society as a whole and how it excludes. But, alternatively, it may suggest that the individual is responsible for integrating themselves *into a society* which is itself fundamentally sound. We shall argue, to the contrary, that it is precisely normal society which creates social exclusion. Third, the discourse of 'inclusion' avoids confronting the question of *justice*: one can be included in any social arrangement in an unjust fashion. These are symptomatic silences of this discourse.

Different terms – poverty, social exclusion, inequality, oppression, injustice – are thus politically loaded. In this book, our view of 'what the problem is' will emerge *gradually* from our history (Part I), social analysis (Part II) and critique of policies (Part III). We thus sometimes use terms precisely, with their distinct meanings; in other places we use them more loosely simply to denote some kind of disadvantage. Our view of the nature and politics disadvantage should be clear by the end of the book.

Whatever words one uses, we have no doubt that there *is* a major problem in contemporary Britain and other developed countries. Differences in income between, for example, the richest and poorest 10 per cent are enormous, and differences in accumulated wealth even larger; in the last 20 years these differences have increased. Large numbers of people work in jobs which are not merely insultingly low-paid but exhausting, unhealthy, worrying, undignified and tyrannical; these suck people's energies and creative capacity out of them. Not only are state benefits low but also claimants are subjected to humiliating forms of surveillance and direction. Many people are excluded from the social norms of commodity consumption, with not only material effects but also severe loss of status. Women, racialised minorities, people with disabilities, young and old people are more likely to experience poverty, which is both produced by and compounds their social and cultural oppression; the latter is also true for lesbians and gay men. Most of the poor live in neighbourhoods with very poor physical environments and services. For a scandalously large number of people there is failure of biological sustenance due to poverty: hunger and malnutrition, living on the street, chronic ill-health, frequent subjection to violence, and, in the case of elderly people, dying of ill-treatment or the cold. All these degrees and forms of deprivation are the cause of deep anguish, often producing low self-respect or self-hatred. They often lead to physically and mentally damaging forms of escape and to mental illness. Any observer with a modicum of humanity must be appalled that such suffering exists in countries of unprecedented high economic output. This book is directed at exploring why this occurs and how it might cease to be the case.

EXPLAINING EXCLUSION: FROM THE BODY TO THE GLOBAL ECONOMY

Our discussion suggests that poverty can be theorised in a variety of ways, and hints that space may form an important element in these explanations. Popular explanations of poverty differ in the *spatial scale* of the phenomena said to account for poverty and in their focus on, respectively, *the economic, the social and the cultural*. From the late eighteenth to the early twentieth century many commentators understood the poor to be a distinct breed, marked by inferior biology or inherent mental disabilities; though now continued by some socio-biologists, particularly racists, this line of argument is now little used. Starting around the same time and continuing to the present, upper- and middle-class opinion has been struck and disturbed by the alien behaviour patterns and culture of the poor. These provide an explanation of poverty with some plausibility: extravagance and failure to save, addiction to drink or drugs, unstable families, failure to study and improve one's skills, reliance on theft – surely these inevitably produce poverty? These vices seem to be perpetuated within poor communities. Rectifying them requires a *moral* crusade. Such an explanation has the advantage of exonerating the economy and focusing attention on 'exceptional' problem localities. It has been taken up in the present day by organic conservatism and, to some extent, by the social inclusion strategy.

A different explanation *does* focus on the economy. Productivist economics (Cole *et al.*, 1983) argues that technological changes and new geographies and modes of organising production have altered the labour market in such a way as to disadvantage some sections of the population, through no fault of their own. The sweated industries in London in the late nineteenth century and de-industrialisation of the north in the twentieth century were sometimes presented in this way. In our day, exclusion from jobs is attributed, variously, to new technology displacing jobs, the 'informational economy' rejecting the poorly educated, or 'globalisation' moving jobs abroad and preventing governments from stopping it. The solution is to equip the poor better for the 'new jobs'. While pointing correctly to some proximate causes of poverty, this explanation averts its gaze from the specifically *capitalist* structures of the economy and society.

In contrast, liberal political economy from Malthus to Friedman focuses on these structures. Thus we have seen how contemporary neoliberalism blames poverty on blockages to markets by states, trade unions and, at times, attitudes which are not 'economically rational'. Marxists, too, focus on capitalist social structures, though in a quite different mode. They argue that capitalist economies and their spatiality produce poverty as the counterpart of wealth and growth, and that class power produces exclusion not only through the economy but within social life and politics. Radical theories also point to forms of power – the oppressions of gender, 'race', sexuality and ability with their varied use of space – as producing social disadvantage and poverty. Our approach uses Marxism and radical theory to construct an explanation of poverty and exclusion and critique others. We emphasise the importance of space (territory, scale, distance) in the exercise of power and resistance to it, and hence in the genesis of poverty.

FOCI AND STRUCTURE OF THE BOOK

This book is concerned with poverty and strategies towards it in the more-developed countries (MDCs). It considers processes common to these countries (excluding the post-1989 Eastern bloc countries), while using Britain as the principal, sustained example. It attempts, however, to problematise the national framework of so much writing on poverty: we explore the impact of international processes on poverty within nations, and the ways in which both poverty and strategy are constructed by relations between scales from the nation to localities to workplaces and homes. Similarly, we consider nested time spans: the history of capitalism, developments in the twentieth century, and the present period of economic and political crisis.

Part I presents a history of dominant ideologies of poverty and the policies and politics associated with them in Britain, from the beginnings of capitalist society to the present. We argue that views of and interventions into poverty have been an integral, and major, part of class relations across the whole society: they are of more than specialist concern. In Chapter 1 these relations are analysed aspatially, while in Chapter 2 we argue that space is integral to them. Chapter

3 presents a brief profile of poverty and exclusion in contemporary Britain. It examines the relation between different aspects or symptoms of poverty, its incidence by social group, and the role of social oppressions.

In Part II we develop an explanation of poverty and exclusion. The focus is on the present period of economic stagnation, but this is set within higher levels of historical abstraction. A key question addressed is the relations between the economy, social life, culture and the state in the production of exclusion. These realms are usually separated in common sense, in academic discourse, and, as we noted above, in most accounts of the causes of poverty. We argue that the principal forms of social power *cross and link* these realms in constructing exclusion. Further, we argue that poverty is not an aberration but rather a normal product of mainstream social processes which also affect the majority of the population.

Part III examines the full range of contemporary strategies towards poverty. At present there is a remarkable degree of consensus around anti-poverty strategy, and we start in Chapter 8 by presenting elements of this consensus and pointing to some immediate problems they face. We argue that beneath the apparent broad agreement, different political currents (seek to) implement these policies in quite different ways. This leads us to examine these currents more deeply. In successive chapters we discuss the anti-poverty strategies of neoliberalism and the extreme right, conservative interventionism, associationalism and socialism. We present the arguments for each strategy, examine the policies used, their supposed mechanisms, and their use of space. We discuss each strategy's actual or potential failings, using the analysis of Part II. We point out how these strategies draw on broad economic and social theories and on particular understandings of space. The critique of each strategy supports the plausibility of the next, and we finish with the socialist strategy that we ourselves believe in.

Within some longer sections, rather than using sub-section headings, we have indicated distinct topics by highlighting their first appearance in bold.

Part I

THE REALITY, IDEOLOGIES AND MANAGEMENT OF POVERTY

1

MANAGEMENT OF THE POOR, IDEOLOGIES OF POVERTY

1.1 THE CREATION OF POVERTY

Strategies towards poverty and the poor in Britain have been bound up with its transformation to a society dominated by capitalist social relations. Over the last five centuries, poverty in Britain has not been caused to any significant extent by the ancient terrors of famine, ecological disaster and plague. Rather, particular social relations of production and social life, which have become more strongly capitalist over time, have created poverty. In this chapter we examine elite views of the poor, the management of poverty, and the conflicts around poverty during these five centuries of capitalism. We shall argue that provision for the poor and management of them has had a central role in the class relations of the society as a whole.

Capitalism has created new types of dependent relationships and new instability in livelihoods. First, the majority of the population has come to depend increasingly on earning a wage. The wage relationship developed in the English countryside from the late Middle Ages, and more rapidly in the eighteenth century with the enclosure (i.e. privatisation) of common land and the growth of commercial agriculture. Many were thereby driven from the land and from the late eighteenth century those evicted were driven to take up wage labour in the new and growing industrial towns. The possibilities for people to live by producing their own food, housing and transport were gradually reduced. These commodities could now be only obtained on the market. More people became dependent on a wage income to buy their basic subsistence, and thus became subject to the discipline of employers, the variability of the wage, and the threat of unemployment. The accumulation of productive capital at one pole created dependency and poverty at the other.

Second, in the long term the average wage has been able to purchase an increasing volume of commodities. This has eroded the skills and availability of equipment needed to produce things such as clothes and furniture and, to some extent, entertainment and the cooking of meals. Moreover, completely new commodities have become an essential part of social life. Those on low incomes are then deprived of an *increasing* volume of consumption, widening their distance from the social norm. Plenty without equality creates poverty.

11

Third, particular industries, and employment as a whole, have become increasingly susceptible to competition, and hence liable to collapse or to relocation. From the industrial revolution, industries have experienced major relocations within Britain; with the rise of advanced industrial rivals from the nineteenth century, and the growth of manufacturing in the Third World in the late twentieth century, there have been major shifts between Britain and other countries. Beginning in the nineteenth century, Britain exported finance on a large scale, and in the last century it has become a major conduit of internationally-mobile finance. These flows, and the monetary policy that accommodates them, have worsened industrial decline. The instability of industries-in-place produced wage cutting, unemployment, and the devaluing of skills. In periods when employment decline was generalised across the national economy, it caused cuts in state spending, in particular on welfare.

Fourth, this increasing role and mobility of finance has impacted on working-class consumption. Housing became increasingly subject to swings in the financial markets. Speculation in land and property impacted on urban uses, often to the detriment of the poor. Old age pensions became increasingly subject to stock market fluctuations.

Fifth, all these developments radically altered the form of the family, gender divisions of labour, and neighbourhood dependencies. The need for wage labour forced people to be more mobile within Britain and beyond, severing neighbourhood ties. It drew in women as well as men (though with periodic reversals as well as advances); this weakened, though did not depose, the ruling norm that women perform the lion's share of domestic and caring work. Family households became more isolated, with poorer households more vulnerable to economic shocks, in turn exacerbating internal conflicts.

Finally, although the spread of capitalist relations to the whole world seems inexorable, the effects have been highly uneven spatially: particular skills become suddenly redundant in one place, whilst there are shortages of skilled labour in another. This has resulted in high levels of both emigration from Britain and immigration into it. Immigrants to Britain since the late nineteenth century have been fodder for, and have sustained, low waged employment, and have been subjected to manifold forms of racism.

The extension and greater intensity of capitalist relations – growth of trade, production and finance, and increasing average commodity consumption – have thus *created* poverty in new ways, and produced increasing economic risk and instability for the majority. Space has been at the heart of this process: there are ever-widening flows of commodities, finance and production, and the continual creation of new patterns of spatially uneven development and instability.

This 500-year long production of poverty has produced an equally long lineage of *policies* towards the poor. The continuity in the causes, forms and consequences of poverty is reflected in a remarkable stability of the ideologies and policies adopted by the elite to manage it. The problems seem constant: women were over-represented among the poor in Tudor times just as they are today. Policies, too, echo down the centuries. Panic over female promiscuity caused James I to legislate to control it and contemporary conservatives to launch their attacks on single mothers. The propertyless, whether they are Elizabethan vagrants or today's asylum seekers, are criminalised. A contemporary proposal to give the poor an acre and a cow (Fairlie, 1996) reprises those of General Booth and William Pitt respectively one and two centuries ago. The principles of contemporary welfare-to-work policies were also those of the progressive reformer Elizabeth Richardson in the mid-nineteenth century: 'there should be a POOR HOUSE for the deserving poor, a WORKHOUSE for those able to work, and a PENAL WORKHOUSE for the idle and thriftless' (Bensham Settlement, 2003).

In this chapter, we examine these recurrent themes in views of the poor and policies towards them. We will argue that this constancy is because of the continuity in the fundamental social relations of Britain. We shall show that all the main historical themes are present today, suggesting their embeddedness in capitalist class relations. Because we wish to emphasise this continuity, we structure the chapter by themes rather than by historical periods.

1.2 OBSCURING THE CAUSES OF POVERTY

Poverty in the world's richest societies is an embarrassment to the elite, who have responded partly through *evasion*. First, poverty may simply be denied. In the nineteenth century, the middle class generally regarded poverty as the *natural* state of the working class: 'the poor are always with us'. In the 1980s, Conservative ministers sometimes maintained that there was no poverty in contemporary Britain because low income people had living standards which were much higher than 50 or 100 years ago or than those of the Third World poor. We offer a critique of this view in section 3.1.

Second, when poverty cannot be ignored, discussion focuses only on its symptoms. Particular manifestations of poverty are seen as *the* problem, such as poor housing and environment, or overuse of drink or drugs. Reform of a limited area can then be undertaken, with predictably meagre results. One symptom is said to be the cause of another, or is conflated with it: well into the twentieth century poverty, lunacy and crime overlapped each other and were treated in similar ways (Marcus, 1969: 43). Moralism can then easily substitute for analysis: late nineteenth-century thought saw the moral and physical weakness of the poor as one; in the contemporary USA and UK, the supposed lack of work ethic of the poor and the decline of the traditional family are conflated (Murray, 1990; Deacon, 1999).

Third, the poor themselves are blamed for their situation. Two favourite targets, constantly vilified over 500 years, are the 'sexually irresponsible' young woman and the out-of-control young man; these produce 'dangerous areas' – dangerous because of sex and violence. These have loomed large in recent discourses concerning the poor – teenage mothers, yobs (Levitas, 1998). These fears have often crystallised in moral panics against the unsettled – vagrants, travellers, the homeless, beggars, asylum seekers. Another constant theme has been the work-shyness of the poor. Work was first linked to morality in the sixteenth century: welfare was to clamp down on the wilfully and inexcusable idle. Populations surplus to business's current demands, a potentially explosive issue, have been vilified as savages (the Highlanders), as obstructions to modernisation (coal mining communities), or as criminals (inner city residents).

1.3 COERCION VERSUS INCORPORATION, SEPARATION VERSUS INTEGRATION

Despite these evasions, active policies to manage the poor and poverty have been implemented since the sixteenth century. The causes of these interventions are complex. First, the poor have been seen by the ruling class as a threat to order, whether because of theft and violence by individuals and gangs, riots, or, most seriously, aggressive political organisation. Second, capital has sometimes sought to mobilise the poor as a real 'reserve army of labour'. This may be done through coercion into work, but may also imply raising the living standards of the poor to make their labour power more useful. Third, the poor and the working class themselves have exerted pressure for the provision of waged work and for better living standards. These sometimes conflicting, sometimes parallel, class pressures have formed poverty policies and ideologies. Each wave of capitalist development since the sixteenth century has reproduced anew these political pressures: the disruption of work and communities has posed problems of individual disorder, new spatial concentrations of workers have organised politically, and capital has sought to channel the unemployed into growth industries and areas.

As already implied, policies for the poor have two modes, coercion and incorporation. Coercion stigmatises and tends to separate the poor from the rest of the working class, whereas incorporation seeks to integrate them. For capital these are both possible tactics to secure an appropriate labour force and contain discontent; for the poor incorporation is clearly preferable. They are always both present but often one is dominant.

There are many forms of **coercion**. There is direct violence by the state using the military, the police and the criminal justice system. Forms of coercion have been most consistently applied to the unemployed. Under the Tudors those unable to survive in the market were put in the houses of correction, the Bridewells, which also served as prisons. Vagrants, criminals and the mentally ill were incarcerated in order to work and thereby to learn to take responsibility for themselves (Melossi and Pavarini, 1981: 14ff). The rise of the industrial town saw the breakdown of the paternalist control of squire and magistrate. The recalcitrant poor were put in the workhouse, imprisoned or transported, and thus *spatially* separated from the honest labourer to avoid 'contamination'. Today, so-called sink estates and prison have the same effect.

Coercive management of labour uses a number of levers. State management of money and finance can be used coercively (Clarke, 1988; Bonefeld *et al.*, 1995). Deflation can create high unemployment, which encourages employers to impose greater work discipline and cut wages and conditions. Capital flows out of the country putting pressure on labour to make concessions. This pressure can be increased by cutting unemployment benefits and tightening eligibility for them. Repression of trade unions aids the process. All these policies were used by governments in the 1920s and 1930s, and again by the Conservatives in the 1980s.

This type of labour relation tends to elicit, and to benefit from, particular working-class cultures – self-denial, self-control, and inner determination. For workers these make hard work more tolerable and meaningful, while for employers they encourage workers to accept work-discipline. In the eighteenth and nineteenth centuries, the churches, civil society and the state created an interlocking set of ideas about work, taking responsibility for oneself and one's family, and about community and nation. Methodism, for instance, gave work a moral dimension as an antidote to idleness, vice and crime (Thompson, 1968: 385ff). These ideas helped to create the 'cage of circumstances' in which the majority grew up (Nairn, 1972). In the late twentieth century this became the language of enterprise and competition, standing on your own feet, and the inevitability of inequality.

Incorporation of the poor, on the other hand, is pursued by support for those without access to wage income, by subsidising elements of basic consumption, and by universal education and public services, all paid for by progressive taxation. There may be attempts to provide jobs for the unemployed through make-work schemes, through reflationary monetary policy, or through national controls over capital flows. Employers may tolerate, and even welcome, trade unions. A working-class voice in government and governance is grudgingly accepted. These have been responses to pressure from the working class, but have also sometimes been supported by enlightened sections of capital: those seeking a skilled, stable work force, and those with the foresight to avoid rebellion (I. Gough, 1982).

In contrast to coercion, this social-democratic approach aims to *integrate* the poor both socially and spatially with the rest of the working class: that is, the span of inequality should be limited; the poor should benefit from statutory employment rights; housing for the poor should be mixed with that of the better off and so on. Most importantly, upward mobility from poverty should be facilitated. Incorporation therefore propagates ideas of common good and one nation. The poor, as much as the rest of the working class, are encouraged to feel that they have dignity and rights.

Over the history of capitalism in Britain, there have been numerous **swings in the emphasis between coercion and incorporation**, and varied **melding** of the two. In the most general terms it has depended on three processes: the cycle of profitability and growth; the balance of demand and supply for labour; and the political confidence of labour and capital respectively; these are related but partially independent. Thus, when profits are low, capital is concerned to impose greater discipline over labour, welcomes unemployment to support this discipline, and is not prepared to shoulder the costs of reproducing the poor; when profits are high business may tolerate greater bargaining power for labour and financial support for the poor. If labour is plentiful – through depression or through immigration domestically or internationally – then capital may use the opportunity to push down wages; if there are labour shortages, capital may seek to mobilise the labour power of the poor and to incorporate them socially. Finally, working-class political pressure for incorporationist strategies can occur not only in times of full employment but also during circumstances, often international, such as revolution and war.

We can perceive these pressures in **the alternations of policy** since the eighteenth century. The first stage of industrialisation between 1780 and 1840 was strongly coercive. Existing working practices were dislocated, creating a large reserve army of labour. Little organised resistance was possible to the appalling conditions for the industrial workforce. Moreover, the ruling class was panicked by the French and American revolutions. Repression occurred through the factory, the army, the workhouse, the criminal justice system, the Combination Laws, and deportation of the 'underclass' to the colonies (Hughes, 1988), extending to stepped-up repression of male homosexuality (Crompton, 1985).

As long as labour supplies were forthcoming and skills were low, worker loyalty and motivation counted for little. Work could be intensified, with no thought about where the next generation of workers was to come from. In the century to 1850 real wages fell by at least 10 per cent, the working day was lengthened. If productivity dropped because of the conditions of work, the working day was lengthened yet further. In addition, the quality of food fell, accidents increased, and fines, high rents and payment in kind reduced workers' income.

From the point of view of capital itself, however, this system reached limits. In the 1850s the supply of labour from the countryside had largely been used up, and, given the terrible conditions, was not being replenished in the cities. The physiological and psychological limits to the lengthening of the working day and intensification of work also began to be evident. Robert Owen had recognised that output per day increased as the working day was *shortened*. Moreover, in some industries productivity could be raised dramatically through the use of machinery, and in others by the use of more skilled and experienced workers. In these industries, brutalisation was counter-productive; employers saw the advantages of raising workers' skills, ensuring their health and welfare, and stabilising their reproduction. There began to be pressure from some sections of business for incorporation of the working class and the poor. Some followed Owen's benign example and built company towns.

Many of the positive changes that have occurred since the 1850s have been partly or mainly caused by reformist **political pressure from the working class**. In the long wave of expansion of the 1850s and 1860s, labour markets became tighter and skilled unions grew, while in the wave of expansion from the 1890s to the First World War the unskilled unions grew rapidly. In response, business conceded union recognition and higher wages; housing, water and sanitation were improved; universal education and state income support were introduced. Poverty came to be pictured as a malaise of laissez-faire social policy (Rose, 1972: 32). The male working-class franchise was granted and the Labour Party formed in the period 1900–18. Another instance of working-class pressure was in 1916–26 when wartime conditions, disillusionment with the slaughter of the war, and the example of the Russian Revolution put the workers' movement on the offensive. This bore fruit in the birth of social work and public housing. Similarly, working-class discontent during the 1930s and the Second World War, the experience of planning, rationing and evacuation during the war, together with the role of the USSR in defeating fascism, produced working-class pressure that fed into the post-war expansion of the welfare state. The ruling class feared working-class demands and accepted that 'modernisation' of welfare was necessary for economic efficiency. Another historically crucial example was the civil rights movement in the USA in the 1960s, and the subsequent uprisings of the black ghettos, to which the response was the Great Society programme against urban poverty.

The incorporation of the poor has, however, **never been free of coercion**. Welfare has enhanced control over the poor by distinguishing between the deserving and the undeserving with white males more naturally deserving than, say, non-white females. From the late nineteenth century, social housing was a means of surveillance and discipline. State benefits have stripped all privacy from individuals and their households and reinforced their internal relations of dependence, such as women on their spouses. Pacts of capital and the state with the working class were framed by international inequalities and the discipline of Britain over the colonies. This was most explicit in the Social Imperialism of Joseph Chamberlain which said to the poor: you will be cared for because you are British and white.

The incorporation of the poor by the post-Second World War welfare regime came to an end with low average profitability, economic stagnation and rising unemployment from the late 1960s. These inclined capital once more to turn to disciplinarian methods, which intensified with the rise of neoliberalism from the 1970s. A highly coercive regime for the poor was constructed through deflation leading to high unemployment, export of capital, and cuts in benefits and public services; their disempowerment was cemented by spectacular defeats of the trade unions. But in the 1990s problems began to appear with this disciplinary regime. Many of the unemployed were not useful to employers because of their skills, attitudes or location, so that their 'human capital' and the state spending on them were wasted. Their individual and sometimes collective lawlessness was perceived as a threat by the rest of the population. The Major and Blair governments have therefore modified and blended neo-liberal coercion with an *element* of incorporation. This project has aimed, above all, to integrate the poor of working age into waged work, thus simultaneously improving business's access to labour, lessening welfare costs, and heading off political instability (section 10.1).

Shifts in the mix of coercion and incorporation of the poor have, then, been powered by cycles in profitability and employment and by conflict between capital and labour. Many of the dilemmas, contradictions and failures of poverty policy lie in these shifts and the consequent need for rapid adjustment of concepts and policies. In the next four sections we examine in more detail the historical continuities and changes in policies towards work, culture, consumption and political engagement.

1.4 PUSHING THE POOR INTO WORK

Capital seeks to maximise the size of its available workforce in order to strengthen its bargaining power against labour – providing that the costs of maintaining the non-employed are not too onerous. It therefore often seeks to make all people of working age into actual or potential workers. Moreover, in contrast to earlier societies, in mature capitalism waged work is the prime form of *social and political* control over the population. We can examine four broad ways in which people have been pushed into waged work since the rise of capitalism.

Creating the need for money

People have been forced into work for money by removing other forms of subsistence. Early capitalism replaced communal forms of subsistence with commodities. After the first enclosures, labourers survived by using common lands for food, fuel, clothing and recreation, but these means of subsistence were taken from them by the loss of gleaning rights, the replacement of annual contracts with casual work, and the marketising of agricultural housing. From the seventeenth century community healers and wise women were suppressed as commercial medicine grew. In the second half of the nineteenth century, leisure became increasingly commercialised, the ability to keep animals for food was curtailed, and the lengthening of the working day left no time for producing subsistence. The need for money-income helped to create the wage-labour force of the nineteenth and twentieth century.

By the twentieth century non-marketed forms of subsistence had shrunk to the unpaid labour of people (mostly women) in their own homes. The policy of state and capital has for the most part switched to *support* for these forms of subsistence so as to reduce the need for state benefits of those unable to live on wages (section 8.7). Nevertheless, capitalist development continues, through new means, to increase the need for money-income. New commodities shape the social life of the majority and thus become necessities, pushing the poor to seek waged work.

Income support and/versus waged work

Since early capitalism, the unemployed of working age have sometimes been able to find some sort of money income from charity or the state. This poses dilemmas for capital. If income support is too much, the incentive to work is apparently undermined; if too little, the workforce is brutalised and may cease to provide an *effective* reserve army of labour.

Early poverty policy was a response to the post-feudal 'free' labourer. Enclosure and the dissolution of the monasteries led to a mass of vagrants who resorted to begging and theft. States responded with laws against begging aimed at forcing the destitute into work. Welfare was centralised and standardised in Britain and France because charity from individuals and the Church had encouraged the able-bodied to beg. The state mandated local administrators to licence those entitled to receive alms and to pay for it through local taxation (Piven and Cloward, 1972). Benefit rates were set well below the lowest wage and below socially accepted subsistence levels to avoid undermining the incentive to work. This is the famous principle of 'less eligibility'. The New Poor Law of 1832, which abolished the more generous Speenhamland system, reaffirmed this principle; it has underlain recent cuts in welfare on both sides of the Atlantic.

From the first, benefits provided to the unemployed responded to *local* labour market conditions: where employers sought new labour, benefits would be reduced. In recent years, the management of the unemployed in the USA and Britain has again moved towards this local management (section 5.9). The unemployed in regions of low labour demand were forced to migrate to where the jobs were, a system recently reintroduced in a number of countries. The spatial scale of management of the unemployed has thus been an important field of contention.

Maintaining the work readiness of the unemployed

The payment of some kind of income to the unemployed may be sufficient to stop them starving but it may not be sufficient to reproduce them as a source of labour power useful for employers. Lack of work experience and long periods of unemployment mean that people lack both work-related skills and habits of waged work; a labour force useful for employers must be cultivated (Cohen, 1991). In localities where work is scarce over decades and where generations of people have lacked formal employment, there are strong motivations to find other ways of subsisting: unregistered casual work for money or as a favour, barter of labour, and theft and illegal industries (Mingione, 1988). Moreover, since the industrial revolution, small groups of people have rejected waged work as alienated and tyrannical, and sought to survive by return to the land and cooperative work in utopian communities (Abrams and McCulloch, 1976).

There is therefore a long history of schemes to convince workers that there is no alternative to wage labour and ensure they remain available for work. Wage subsidies – from the Speenhamland system to contemporary tax credits – lure the unemployed into low paid work. Make-work projects, such as the slave camps of the 1930s (Hannington, 1937) and the contemporary imposition of community work through workfare systems, teach the unemployed the disciplines of wage labour; in recent years the 'social economy' has served the same purpose. When waged work is scarce, self-employment may be promoted to maintain work habits, as in the 1980s, or sections of the population may be removed from the labour force, as in the expansion of higher education since the 1970s. Training schemes and mentoring for the long-term unemployed attempt to increase flexibility. Carrots and sticks are thus used to retain the unemployed as a real labour reserve.

Even when people become employed, interventions may be necessary to acculturate them to work rhythms required by the employer. Wageworkers from the sixteenth century onwards have had to be punished for taking off 'Saint Monday' and celebrating lengthy festivals (Gutman, 1976). Early factory workers in Britain had to be socialised into a completely new notion of time, governed by the machine rather than nature (Thompson, 1967); this occurred also with the millions who migrated to the USA in the nineteenth century and early twentieth century from pre-industrial cultures. Nowadays, employers can sack people on work-experience and trainee schemes if they fail to show sufficient discipline (Peck and Theodore, 2001).

Work as social discipline

Political and social control of the poor by the elite is a particular problem within capitalism. Feudalism required all workers to have a named master to take responsibility for them. In the late medieval and early modern period, the free labourer caused increasing problems of control as we have seen; but many remained subjected to surveillance and control by squire, priest and magistrate and were under pressure from neighbours to abide by traditional norms. With industrial capitalism the free waged labourer becomes the norm. Their concentration in the anonymous city caused the disintegration of traditional forms of control, a serious problem for capital suggested by the fact that it was a central preoccupation of classical sociology. In reaction, some large industrial employers constructed paternalistic systems of control over their localities through providing welfare and recreational facilities, particularly in company towns (Brandes, 1970). But these are hard to sustain, and even harder to generalise, in developed capitalism: competition may require relocation, as with the withdrawal of paternalistic Pilkingtons from St Helens and ICI from Teesside in recent decades (Beynon et al., 1989); residents of most neighbourhoods work for diverse employers; workers are spatially mobile. In consequence, the state takes on a growing role as guarantor of social order through welfare and policing. However, the rights of the independent wage labourer, property owner and citizen limit the state's interventions.

There is, nevertheless, one powerful form of immediate social and political control in mature capitalism: waged labour. If individual workers are not socially respectable in behaviour and appearance, or if they develop a reputation as rebellious, they will not get work. If residents of a locality develop a reputation as unruly or radical, the locality will not get investment. Moreover, once inside the workplace, the employer subjects the employee to direct control over (potentially) their every action in order to ensure surplus labour is performed (Marx, 1970). Waged labour then requires obedience and self-discipline that can then inform behaviour outside the workplace. Thus, the economic aims of employers function, as an unintentional consequence, to impose social and political discipline (Burawoy, 1985).

This means that long-term unemployment is a *social and political* problem for capital. In communities without jobs over long periods the 'normal disciplines' tend to break down. If a 17-year-old youth knows he will never work he has no reason not to misbehave, and in a gang he has no reason not to riot. Forcing the unemployed into waged work is therefore motivated not only by employers' demands for a larger workforce but also by a wish for social order. That is why such programmes sometimes operate when there are no jobs available. New Labour's wish to re-impose social discipline on the poor partly explains its enthusiasm for welfare-to-work.

1.5 INVENTING AND TACKLING THE UNDERCLASS AND THE CULTURE OF POVERTY

Social discipline is, indeed, one of the most durable themes in discourses of poverty. We have seen how elite discourse often blames the poor themselves for their condition (section 1.2). The latter has sometimes been attributed to biological or mental defects. But more common, from the sixteenth century to the present, has been the notion of a dysfunctional *culture* of the poor, which perpetuates their poverty, particularly by their failure to obtain waged work. These *disorderly* poor constitute an 'underclass'. Inadequate socialisation from poor mothers and criminal fathers is a recurrent theme: the early nineteenth century saw panics concerning inadequate mothering in the new industrial towns (Davin, 1978); current conservative discourse on poverty among Afro-Caribbeans focuses on absent fathers (Murray, 1990). The poor are seen as hedonistic and in search of instant gratification, and as destroying their labour power through drink and drugs. Crime and disorderly behaviour in poor communities is seen as both a cause and a product of a disordered society. The society of the poor is pictured either as lacking in social cohesion, or as atomised and fragmented, or as not a society at all (Forrest and Kearns, 2001).

The culture of poverty theory obstructs a real understanding of poverty. Poverty is not caused by the poor themselves but is endemic to capitalism. The

Figure 1.1 Frontispiece to General Booth's *In Darkest England, and the Way Out*, 1890. A holistic approach to poverty using city, countryside and Empire. The title of Booth's manifesto alludes to Africa (cf. section 2.2). The Salvation Army is now second only to the state as a welfare provider.

poor have their own forms of social order and the culture of the poor is a rational adaptation to their circumstances. The shakiness of the notion of the underclass is reflected in its elastic definition. The underclass can include any group of whom the respectable public disapproves, exemplified by General Booth's famous illustration of those drowning in the sea of despair (Figure 1.1): strikers and anarchists share the water with divorcees, harlots, murderers, the homeless and perjurers (Booth, 1890). Despite this vagueness, many of the welfare policies we examined in section 1.3 have been directed at 'improving' the orderliness and daily behaviours of the poor, what Mann (1992) calls 'civilising the yobs'. In nineteenth-century Britain, the university settlements, missions and armies of middle-class women armed with religious tracts thrust themselves upon the poor, with incontrovertible evidence of why they should change their ways; in the twentieth century this role has been taken up by state-organised social work and council housing. Both have aimed at social order and political docility (Ambrose, 1994).

Many interventions have sought to *separate* the respectable poor from the underclass. In the late nineteenth century, private charity sought out those on the borders of pauperism who were willing to improve themselves. To qualify for assistance the poor were carefully vetted in order to discover their motivations and values; thus improved housing was made available only to the respectable, and came with draconian rule books stipulating good behaviour (Stedman Jones, 1971). Surveillance has always been an important part of welfare in order to divide and rule: rewarding the respectable, reforming the pauper and punishing those who will not try. Yet the boundaries between 'rough and respectable' have always been fluid because of the vicissitudes of capitalist accumulation and the ease of dropping into poverty.

A longstanding focus of intervention has been the fertility of the poor, directed at improving the upbringing of children. In the nineteenth and early twentieth centuries the eugenics movement sought to limit the 'uncontrolled breeding' of the poor in favour of the reproduction of 'better stock'. Late nineteenth-century town planning proposed housing improvements with the idea that less overcrowding would reduce the birth rate of the poor (Hebbert, 1980: 13). Many women classified as mentally ill, 'simple', 'immoral' or 'promiscuous' were forcibly sterilised (Wilson, 1977). Direct limiting of the poor's fertility now only happens in the Third World. But the teenage mother and the mythical young woman 'who gets pregnant to jump the housing queue' are still with us.

Racialised groups have been subject to the most comprehensive and coercive forms of 'improvement'. Irish immigrants in the nineteenth-century USA were subjected to custodial facilities, compulsory schooling and powers to remove children from their families (Schlossman, 1974). From the early twentieth century to the 1960s Australian state policy was to eliminate aboriginal society by integration, on the grounds that aboriginal people were disorderly and opposed to wage labour; the forcible removal of children to be placed in white families was the result (Moses, 2004). Such comprehensive attacks are often justified by moral panics. Early last century, for instance, cocaine was demonised in order to legitimate repressive measures against blacks in the southern states of America (Pearson, 1975: 58).

The present period has seen the revival of notions of the disorderly underclass, but presents different explanations of it. Conservatives focus their blame on the decline of the family, and see modernity and commercial culture as responsible for individualism and selfishness, eroded communities, mindless consumption and hedonism (Scruton, 2001). Neoliberals castigate over-generous welfare, which removes the incentive to get a job and better oneself. Conservative interventionists share these views, but see the underclass as requiring *social* reform rather than simply moral exhortation or economic incentives.

1.6 INTERVENING IN CONSUMPTION

Much policy towards poverty has consisted in interventions to cut the price or improve the quality of the basic consumption goods of the poor. Part of the pressure for this has come from the working class. It has also sometimes come from employers keen to reduce wages without endangering reproduction of their workforce, for instance by reducing housing costs. Both workers and employers have sometimes sought to reduce prices of housing and local transport in order to increase access to jobs.

Improvement in the commodities consumed by the poor requires either regulation of producers or their ownership by the state. Over the period from the late nineteenth century until the 1950s a number of utilities – water, gas, electricity, rail, buses – were taken out of private ownership, partly in order to improve services for the poorest consumers; health care followed suit in 1947. Rented housing was first seriously regulated during the First World War. Town planning from the late nineteenth century was aimed to reduce costs and improve the quality of the built environment for the poor. Sometimes governments attempted to control subsistence costs across the board: the Conservative government of 1970–4 offered to control basic prices and rents, though in exchange for wage restraint: capital had made the connection between wages and workers' subsistence costs.

The state's control over consumption goods may, however, be resisted or reversed by capital unwilling to lose the associated profits. The complete regulation of the quality of food and drink, for instance, has always been ineffective against the politically powerful farmers and agribusiness. Since the 1970s the regulation and state ownership of subsistence industries has been put into reverse by neoliberalism. The result has in most cases been increased costs for the poor. Alarmed by the effects of this on the reproduction of the poor, conservative interventionists have dimly realised the logic of the old arguments for social control of basic consumption, and are experimenting with non-market, non-state forms of provision. Thus, the deep tensions around working-class subsistence consumption continue to generate policy somersaults.

1.7 POLITICAL INCLUSION AND EXCLUSION

The achievement of universal suffrage formed a part of moves to incorporate the working class from the late nineteenth century (section 1.3). But the elite has never lost its fear that enfranchising the working class would endanger capitalist class control. In consequence over the last century the vote has yielded only limited influence for the working class and even less for the poor (Vincent, 1991). A number of processes have been at work, including shifts in spatial scale.

First, the ruling class has often changed the rules when lower levels of government have fallen into radical working-class hands. In 1834 Boards of Guardians, which administered the Poor Law were established. Long run by local elites, some succumbed to democratic control. This potentially gave the working class and poor considerable power, since the propertied subsidised the property-less at a rate determined locally. This threat was met by transferring responsibility from the locality to central government: 'It . . . took just a decade-and-a-half of full participation by the poor in their own relief to remove the element of local responsibility which had lasted for more than three centuries' (Vincent, 1991: 62). Administered centrally, the pattern of redistribution was changed to be *within* the working class, from those in work to the unemployed and from smaller to larger families. Similarly, from the 1920s to 1940s the local authorities had key sectors for anti-poverty policy transferred to central government – power, transport, police, health and, to a large extent, housing. Even then, the local authorities have been regarded with distrust by the ruling class as possible foci of working-class influence, as was demonstrated by Lansbury in Poplar. Their actions have to fall under powers specifically bestowed by parliament, they are subject to detailed central government direction, and they have received (an increasing) majority of their funding from the centre, as their powers to raise taxes locally have been restricted. When the Metropolitan (city-region) councils tried to implement left policies in the 1980s as explicit alternatives to Thatcherism, they were simply abolished.

Second, from the nineteenth century the dominant politics of the trade unions and the Labour Party, most strongly expressed by their leaderships, has been rooted in Puritanism, self-help and the traditions of the male craft worker. They have seen their legitimacy as flowing from parliamentary democracy with its established structures, and they have been suspicious of direct action such as the National Unemployed Worker's Movement in the interwar period; they have thus failed to develop a properly Jacobin tradition. In the early twentieth century the leadership of the labour movement was suspicious of the poor, expecting that they would indulge in shirking and welfare fraud, and therefore rejected high taxation to finance benefits (Vincent, 1991).

Due to these conservative pressures, the state welfare services, which were the most important aim of working-class enfranchisement, have been run in a highly bureaucratic fashion. This contrasts with the democracy, cooperation and mutuality that had organised welfare for the skilled working class in the mid-nineteenth century. The poor have been prevented from controlling the administration of benefits and public services. The state, having initially appeared as a saviour of the poor's living standards, turned out to be remote and controlling.

Nevertheless, the post-Second World War welfare state for a time created a sense of trust and unity. Universal welfare – social insurance, the National Health Service, the Education Act, strong land use planning, nationalisation of basic industries, coupled with a pledge of full employment – for the first time gave working-class people a sense of inclusion in national politics. The *national* scope of this settlement, although earned at the expense of local auton-omy, strengthened its appeal by promising equality irrespective of locality and creating a powerful national constituency for welfare. But the tragedy of this post-war settlement was that there was no holistic vision. The nationalised industries failed to change relations in production, education ignored its role in changing class boundaries, the NHS failed to see the relevance of poverty to ill health, and land use planning perpetuated the power of property capital and the well-off (Mackintosh, 1987). Political incorporation, therefore, provided benefits but within a framework that institutionalised inequality and class power.

The political incorporation of the poor during the post-war period was sharply reversed by neoliberalism. Political centres, which expressed the poor's aspira-tions, were suppressed. But this led to the poor using other means of political expression – refusal to pay the poll tax, riots. One million urban young men disappeared from the electoral register. This was seen as a threat by conservative interventionists, who have attempted to open new forms of democratic partici-pation to the poor through partnerships of their communities with the local state. These swings in the balance of political incorporation and coercion can be traced in changing conceptions of *citizenship* (Faulks, 1998: 5). During the post-war period, citizenship implied civil and social rights independent of income; citizens of the state, however poor, could expect equal treatment from it. Neoliberalism proclaimed that these rights no longer existed. The poor would have to fend for themselves in a 'free' labour market rather than rely on the state. They should not rely on the state as a provider since the state could not solve their problems but was actually a *cause* of poverty. Neoliberalism thus tended to expunge the idea of citizenship. The turn to CI, however, has partially rehabilitated the notion of citizenship. It does not return to citizenship as universal rights, but rather urges the poor to become active citizens to fight for, or create, their community's welfare. Swings between coercion and incorporation of the poor thus have an important political component.

1.8 CONTINUITY, CYCLES AND DILEMMAS

There have, then, been remarkable *continuities* over 500 years of British capitalism in ideologies concerning the poor and in policies towards poverty.

These exist because of the durability of the processes of producing poverty (section 1.1), and because of the durability of the fundamental class, patriarchal and 'racial' relations within which poverty policy is set.

These relations are, however, conflictual and contradictory. Capitalists need to keep wages and benefits down, but they also need to employ a reasonably fit and skilled workforce. They must seek to prevent the growth of strong collectivities and militant political action by the poor, but they also need a more-or-less willing labour force. The development of capitalism has produced wage earners as the largest single social group and they have achieved considerable economic and political leverage, periodically creating considerable pressure to improve the poor's conditions. Incorporation may become excessively expensive for capital and undermine its control, but coercion may damage productivity and lead to rebellions. These conflicting pressures produce significant *variations* in policy affecting the poor, involving the constant re-balancing of coercion and incorporation. By the same token, management of the poor is not a marginal aspect of politics but central to the development of capitalist social relations. The next chapter continues the discussion by exploring how poverty policies use space.

2

SPACE AND
THE MANAGEMENT
OF THE POOR

2.1 POVERTY POLICY USES SPACE

In the last chapter at various points we referred to the role of space in poverty policy in capitalist Britain; in this chapter we consider this role more systematically. Space is central to poverty policy because the geography of capitalism is inherently dynamic. Capital constantly seeks out new labour forces, production linkages, final markets and political conditions (section 1.1). In doing so it frees itself from the constraints of any one place. As capital moves, its new absences and fresh presences change places, create new ones, and produce new spatial patterns of inclusion and exclusion. In consequence, space – by which we mean distance, place and scale – has been a *constitutive element* in policies and ideologies of exclusion.

This is so in two fashions (cf. Walker, 1978). First, policy and discourse have rested on **contrasts between 'poor' and 'normal' places**. Poverty is associated with particular, deviant places, contrasted with normal, unproblematic places from which this gaze is directed. This contrast is made at all scales: the civilised mother country is contrasted with the primitive colonies; the growth region with the depressed; the traditional, orderly countryside with the city of fragmentation and anomie; the respectable suburb with the disorderly inner city neighbourhood. Good versus bad places has been a consistent trope of poverty policy.

The second mode in which space has entered poverty policy is in **managing flows of capital and labour between territories**. Again, this occurs at every spatial scale. Many policies have aimed to give spatial freedom to markets by **enhancing spatial flows**.

- Promoting *flows of capital* has been an important strategy. Inward investment to a poor locality or region or to Britain as a whole has been stimulated. Alternatively, capital has been enabled to flow out of problem areas, increasing pressure on business and workers to restructure, and thus eventually increasing the profitability of the area.
- Policy has often sought to promote *flows of labour*. The Empire was used to absorb Britain's surplus population. At other periods the colonial poor have

been encouraged to come to Britain to relieve its (low paid) labour shortages. The poor are moved from regions with high unemployment to growth regions. The inner-urban poor are moved to the suburbs in the belief that this will solve their poverty.

On the other hand, however, many strategies against poverty have seen free spatial flows as the problem, and have attempted to regulate territories, manage markets, and **reduce disruptive flows**.

- Policies have attempted to *prevent flows of capital* out of an area to stabilise investment and stimulate growth within it. Examples are Keynesian policies for the national economy, and organisation of a locally-networked and contained local economy.
- Other policies have sought to *prevent flows of labour*. Prevention of international immigration to Britain has been justified as reserving resources for the indigenous poor. Management of the urban poor has often aimed to contain them within their existing neighbourhoods to prevent them disrupting or contaminating 'normal' ways of life.

Poverty policy has thus always been, integrally, policy for managing the relations *within* 'poor areas' and *between* these and the 'prosperous areas'. These areas are conceived at **all spatial scales**. In the rest of the chapter we examine these relations starting at the largest scales and moving to the smaller. Policy is not hermetically contained within a particular scale, however; on the contrary, it typically *relates and contrasts* the different scales (Howitt, 1993; K. Cox, 1998; Brenner, 2001; Gough, 2004b); the sections below are distinguished accordingly.

We shall see that both incorporation and coercion of the poor have used space in these ways. So, too, have socialist and radical organisations against poverty, which are considered in Chapter 12.

2.2 THE NATION AND THE WORLD

International relations have had profound impacts on attitudes and strategy towards poverty within Britain. We examine four overlapping aspects:

The external and internal colony: similarity between the poor in Britain and the colonies

In the nineteenth century there was interplay between the images of the subject peoples created by imperialism and those applied to the domestic underclass. Parallels were drawn between subjugated 'races' and subjugated classes; internal and external repression cross-fertilised each other with stereotypes that have become embedded in British culture. Colonial peoples and the domestic working class were regarded as biologically inferior, naturally indolent, dependent and without enterprise. Both lived in areas 'of which little was known' and offered unquantifiable threats; images of 'the Dark Continent' were transposed to poor areas of the cities. Forms of colonial rule were then applied to the governance of the domestic poor. Dominant relations of the upper class to the poor, veering between indifference and repression, were nurtured by the violence and economic ruthlessness of imperial rule (Davis, 2001). The brutality of the treatment of the Irish in the nineteenth century by the British ruling class, for example, was congruent with its treatment of the indigenous and Irish working class in Britain (Curtis, 1984). The army was used as security against both colonies and insurrection at home. 'Missions' and 'settlements' set out to instil discipline, especially wage discipline, in the domestic poor.

This cross-fertilisation is still active. The dominant strategy against poverty in the Third World is now focused on family production, the informal economy, self-employment, community development and self-help; these old methods of survival are currently regarded as practical and resilient. So long as they are complementary to capitalism, they are supported by governments and NGOs in order to build the social capital of the poor: micro-funding for micro-enterprises, in particular for women, is granted; trading cooperatives are encouraged; community organisations using unpaid labour to provide basic services (sewers, fresh milk supplies, and so on) are supported; credit unions fostered; planning and land ownership regulations to support self-build housing are promoted. In the last two decades these policies have been transferred to the First World. Just as the discrediting of Third World statist modernisation strategies gave plausibility to small scale, informal solutions, so the discrediting of social democratic welfare has played the same role in the First World. This links have been made by varied political currents including neoliberals (World Bank, 2005), conservatives, associationalists (Amin, 1999) and left libertarians (Ward, 1976). In both the Third and First World, community and small enterprise are presented as a means of empowering the poor.

Some contemporary political discourses see social practices of the Third World poor as *superior* to those in the developed countries. Neoliberalism seeks to bring Third World wage norms and labour discipline to the developed countries. Though inhibited by popular opposition, it has promoted immigration from the Third World in order to swell the workforce for low paid, intense, dangerous and unhealthy work. In extreme cases the relation between First and Third Worlds may be locally inverted: Malaysia has technically-advanced clothing factories while London still has sweat shops; a Pakistani village sends care parcels to a village in northeast England (Danziger, 1997). Third World models of regulation have been promoted. Peter Hall first proposed Enterprise Zones in the 1970s as creating 'mini-Hong Kongs' (Shutt, 1984). The new governance through partnerships has traits that are pronounced in many Third World countries: key decisions are made outside the formal legislative process; there are corporatist bargaining and 'pork barrelling', and important roles are reserved for city fathers or godfathers. These violations of democracy are legitimated as 'getting things done'. In a softer mode, conservative interventionists promote family responsibilities and dependence, parental discipline and religious faith, which are seen as stronger in the Third World and among immigrant communities in Britain than among whites. These social structures are seen as a basis for escaping poverty.

The domestic poor in imperial projects: contrasting the poor in Britain and the colonies

On the other hand, the tradition of social imperialism emphasises *difference* between the colonies and the mother country. Part of the profits from the Empire was returned to the elite of the domestic working class. This gave a sense of superiority over colonial peoples, and has been a material support for jingoism. Not only was this an effective means of incorporation, but also it socialised workers into their imperial tasks and created that bedrock of British politics, the conservative, skilled, male worker. But this inclusion was premised on the exclusion of the unskilled, women and immigrants, as well as workers in the Third World.

Neoliberalism erodes this politics through its hostility to well-paid industrial workers and its wish to reduce their conditions towards those of the Third World. But, contradictorily, contemporary neoliberal imperialism (Harvey, 2003) legitimates itself through demonisation of Third World peoples, thus drawing on, and reinforcing, 'racial' pride and reinforcing racist exclusions in the First World. These serve, as in the nineteenth century, to cement relations between capital and the white working class, now strained by neoliberal austerity.

Economic competition of Britain with the rest of the world

Since the mid-nineteenth century, production in Britain has faced ever-increasing competition from other advanced countries. This has had profound implications for poverty creation and alleviation. These have, however, been mediated by political strategy. International competition has been met by two broad responses: a liberal strategy of holding down wages, intensifying work and deskilling, and a social democratic strategy of product and process innovation based on cooperation with a relatively skilled labour force. These have been linked to, respectively, coercion and incorporation of the poor (section 1.3).

These impacts of competition can be traced historically. Pressures towards **incorporation** of the working class and the poor became stronger in the late nineteenth century as the economic challenge from Germany and the USA became more threatening and as imperial competition in the colonial world intensified. The slums came to be seen as the rotten heart of the Empire in their lack of work ethic, immorality and failure to produce suitable army recruits. The defeats of the Boer war initiated an intense debate over how to improve the culture and reproduction of the working class (Doyal, 1979: 161).

The international context also fed into the construction of the post-Second World War welfare state. The war had been, amongst other things, one of neo-imperial competition for Eastern Europe and East Asia (Mandel, 1986). Its privations produced intense pressure from the working class in Britain for better conditions and opposition to military-coercive politics at home and abroad. This combined with the imposition of free trade by the USA, using its new position as the dominant country. Free trade even applied to Britain's colonies. Given the threat of working-class discontent, this increased competitive pressure on the British economy was met by a social democratic response.

However, at other times international competition has been met by liberal measures, exacerbating poverty and **coercion** of the poor. The decisive weakening of Britain's power by the First World War and the economic crisis of 1920s caused the elite to resort to economic discipline: there was a return to the Gold Standard and tight money, aggression against the trade unions (of which the 1926 General Strike was symptomatic), and direct attacks on the living standards of the poor such as the cutting of the dole by the Labour government in 1931. This coercion owed much to the dominance of financial over industrial capital in Britain – itself a product of Empire (Fine and Harris, 1985) – since financial capital does not have to manage workers or elicit cooperation from them.

The return of liberal policy as 'neoliberalism' in the 1970s was also powered by international competition and economic flows. Intensified international competition was said to necessitate austerity and attacks on the trade unions. At the same time, capital was to be freed to leave Britain so that it could locate in places (and industries) with higher profit rates. Flows of capital between developed countries increased; much manufacturing capital relocated from the First to the Third World, intensifying downward pressure on wages, benefits and welfare in Britain (Green, 1989; Fagan, 1991; Cox, 1997). Increasing poverty and coercion of the poor were thus justified by two, apparently contradictory aims, to increase British competitiveness *against* competitor countries and to free British capital to relocate to them.

The international politics of British capital has thus profoundly affected policy towards the poor, whether cooperative or coercive.

Emigration, immigration and 'racial' difference

Britain's imperial role – extending to the present – has produced significant emigration and immigration, with direct implications for poverty policy.

Through the eighteenth and nineteenth centuries, the Empire enabled export of the 'surplus' population as settlers, convicts or military personnel. The immediate effect was to remove a potential source of political trouble. This in turn shaped the whole political economy of the period: shedding of labour by enclosure, mechanisation and sectoral decline could proceed because a large proportion of the resulting unemployed, particularly the most rebellious, could be translocated (Prebble, 1969). However, the successful imprinting of capitalist relations on colonies and their independence has removed this safety valve.

During the twentieth century capital in the advanced countries has used the reverse flow, immigration from colonial and neo-colonial countries to expand the workforce for poor jobs and maintain discipline over the indigenous workforce through intensified competition. Upton Sinclair's (1965) account of mass migration from Eastern Europe to Chicago's meat packing industry exposes the basic mechanisms: the mass of cheap labour, their exploitation in the labour, housing and product markets, their politicisation and rapid exhaustion, and their replacement by more 'greenhorns' willing to work for even less.

There are, however, different ways in which immigrant groups can be managed and can manage their place, which can be found across the developed countries. On the one hand, capital seeks to erode cultural differences so as to homogenise the workforce – the 'melting pot' ideal – and thus intensify competition within it. Minority ethnic groups tend to have strong community ties that often translate into political or labour resistance and thus improve welfare and employment conditions generally. But this can lead to a degree of working-class political unity that can threaten business interests. This direction was present, for example, in Australia from the 1950s to the 1990s.

On the other hand, immigrant communities may seek to preserve their community ties and distinct culture as a means of support; employers may use labour of different ethnicity in quite different ways; and business as a whole may see political advantages in perpetuating differences between immigrant groups and the indigenous population. The latter, informed by imperial attitudes, may also exercise racist exclusion where they can. Immigrant and minority ethnic groups thus become strongly socially excluded. The outcome of this tension between cooperation and rivalry between racialised groups is partly conditioned by the overall political-economic context. Social democracy tended to promote incorporation of minority ethnic groups; neoliberalism, with its deepening of competition within the working class, is tending to exacerbate antagonisms (Delanty, 2000).

Through these varied processes, then, the international relations of Britain have profoundly affected policy towards exclusion within it.

2.3 THE NATION, THE LOCALITY AND THE INDIVIDUAL

Since the late nineteenth century, anti-poverty policy and welfare services in Britain have been mainly designed and financed by the *national* state, and have been intended to be uniform across the national territory. This is because Britain has been essentially a single space for class organisations (business, labour, and parties linked to them) and thus for the class conflicts which determine welfare policies. The national state progressively centralised income support, welfare services and infrastructures during the nineteenth and twentieth centuries. This was partly in response to local provision controlled by elites being seen, variously, as too generous or too mean, and partly a response to working-class control being seen as too strong at the local level (see section 1.7). It was also because of the increasing cost of welfare: the national state was unwilling to grant local authorities the necessary taxation powers lest they be used 'unwisely', and local tax bases are so variable that they would result in highly unequal welfare.

Nevertheless, local authorities have retained an important role in administering some public services, and have had a limited control over them quantitatively and qualitatively. This limited autonomy has been used variously for right-wing or leftwing aims depending on local class pressures. Thus some rightwing councils have provided little or no public housing and poor social services, while some leftwing councils have defied central government to attempt to spend more to meet working-class needs (Poplar in the 1920s, Clay Cross in the 1970s, Liverpool and Lambeth in the 1980s). Public services for the socially excluded have thus varied considerably by locality.

Since the 1970s central governments have further centralised control over welfare in order better to mould it to neoliberal precepts. But paradoxically, differences in welfare by locality and neighbourhood have tended to *increase*. This is because as the tax bases of council areas have diverged more, so inequalities between neighbourhoods within them have grown (section 3.6), and welfare has become increasingly tied to employment and economic competitiveness (section 8.4).

A further scalar paradox of welfare is that both locally- and centrally-controlled services have *individualised* the beneficiaries. People have been atomised in dealing with the welfare state, confronting a bureaucracy as individuals. The state has given very limited control to residents' or neighbourhood associations, and has generally discouraged control by workers', women's or minority-ethnic organisations; solidarity has been designed out of the system. Social work, for example, deals with the individual or family and implies that they can find solutions to their problems in isolation (Doyal, 1979). New Labour has adopted the Tory government's Citizens' Charter of 1992, which reformulates citizenship as the customer's right to certain services. This individualisation keeps welfare subservient to the fundamental relation of capitalist societies, the individual wage, and reduces the danger of collective pressure from the poor and their politicisation. We see here, again, that the *articulation* of different spatial scales – nation, locality, home – has been central to the politics of poverty.

2.4 THE NATION AND ITS REGIONS

Since the industrialisation of agriculture and the industrial revolution, uneven development between the regions has been endemic to Britain (section 1.1) and poverty policy has been predicated upon it. Two alternative strategies have been used, with contrasted class relations. First, the movement of the unemployed and poor from depressed to growth regions has been encouraged. This caters to the choices and immediate demands of firms. It accentuates uneven growth. Second, capital has been encouraged to invest in the poorer regions. This allows the poor to remain in their existing social milieux, and it involves some constraints on firms' choices. But it can also benefit business as a whole by using sunk investment in infrastructures and labour power in the depressed regions, exploiting their cheapness, and by reducing congestion and inflation in the growth areas. This second option, too, has consequently sometimes received support from business.

In the eighteenth and nineteenth centuries there were massive movements of labour from country to towns and from agricultural to growing industrial regions. These flows were encouraged by public policy. Workhouses sent their paupers long distances to industrial centres in order to lessen their burden on the rates and feed industry's demand for unskilled labour. Because the city was considered the place of opportunity, population movement was considered the only solution to rural poverty. Encouraging people to move required a revolution in policy because pre-capitalist society had been built on spatially- and socially-tied labour; thus relief could be given only in people's parish of birth. The social relations of housing also had to change radically to provision not tied to individual employers, requiring the creation of a land market and its associated financial infrastructure.

However, from the 1930s to the 1980s regional policy switched emphasis towards encouraging the movement of capital rather than labour. This was partly because of the strength of community and family ties. In addition, the switch during the twentieth century, from private renting to owner occupation and council housing, introduced new forms of spatial friction in housing. Council housing failed to create an effective transfer system and owner occupation was unavailable to the poor. (Nevertheless, there were those who worried that growing owner-occupation would inhibit labour mobility as a whole.) Both of these factors contrasted with the USA and Australia with their high inter-regional mobility of labour. Pressure from the working class from the 1920s to the 1940s was important in initiating regional policy focused on financial incentives to firms to locate in the depressed regions, supplemented by some restrictions on investment in the growth regions. This regional policy was an important component of post-war social democratic anti-poverty policy with its commitment to full employment.

Since the 1980s, however, the strategy of encouraging capital movement between regions has been both scaled down *and* changed in its social-economic and spatial complexion. Incentives to firms have not created equal regional prosperity, and the limited convergence of the boom period went into reverse from the 1970s (Hudson and Williams, 1995). Financial incentives tied to location have been widely denounced as crude, cost-ineffective and statist, and they have been reduced, although not abolished. Regional policy for investment has moved towards changing the *qualitative* structures of both industries and workforce in the lagging regions. Depressed regions are portrayed as having an inappropriate *culture* for the 'post-industrial' economy due to the former paternalistic domination by large firms and nationalised industries, leading to lack of enterprise and 'flexibility'. This view is a reworking of the culture of poverty thesis (section 1.5) applied to whole regions rather than neighbourhoods. Policy thus seeks to change attitudes to re-skilling, self-employment, welfare and gender roles. Industry is encouraged to form inter-firm networks and intensify knowledge transfer and innovation. These interventions are to come from *within* the region rather than from central government: the region itself is to become an 'actor'.

Alongside this change in regional industrial policy, neoliberalism has revived the idea that the unemployed should move to growth regions: industry minister Norman Tebbit notoriously told the unemployed to 'get on your bike'. But housing costs in the prosperous south have become prohibitive for low-paid workers and the stock of social housing has been halved. The strategy of inter-regional mobility has, as in the industrial revolution, run up against the social organisation of housing.

2.5 THE CITY, 'RURALITY' AND THE SUBURBS

Poverty policy has involved not only flows between regions but movements within cities and towns. Since the late nineteenth century, the key solution at this scale has been suburbanisation. The poor in long-established inner neighbourhoods were moved to newly built outer areas of the city ('the suburbs') or to new freestanding towns. This has been seen as helping to overcome poverty in four distinct ways. First, the new housing has been assumed to be physically superior to the old – sounder construction and more floor space, 'proper drains', and more external spaces in gardens, wider roads, parks and playing fields, and better air, thus combining the physical advantages of the countryside with the convenience of the city (Reade, 1987: 37). Second, the new housing was to be occupied by nuclear families, and its layout and spaciousness encourage a more privatised, home-based way of life. This helped to erode the 'culture of poverty' of the old working-class neighbourhoods by breaking the community ties that are still thought to create this culture. The clearer private space of the home has been intended to promote individualism, self-interest and self-improvement, the classic petty-bourgeois values. Third, the suburbs and new towns have been intended to mix the middle and working classes. The old working-class areas perpetuated the culture of poverty by being single-class; they were clearly demarcated from (if often proximate to) middle-class areas of the inner city. The suburbs were to include, and mix, the broad middle of the social spectrum – 'Middle England' – and promote its solid values to the (formerly) poor (Fishman, 1987). Finally, at least in some versions of the suburban utopia, jobs were to be closer at hand than they had been for many inner city residents. The old inner city was seen as a dangerous mix of types of employment, eroding links between employer and residence. Suburbs could potentially re-establish this connection for factory and, later, office work.

In British ideologies of suburbia the physical and social advantages of the suburb are conflated with each other: the poor will be improved by fresh air and greenery; they will regain elements of the pre-industrial rural life, including strong families and a proper sense of social deference. This vision has been fuelled by the characteristically English revulsion against the large industrial city – a fear of the working class displaced into a spatial fetish – and the corresponding idealisation of (largely mythical) 'traditional country life' (Williams, 1975). The British ideology of suburbia has been virulently environmental-determinist.

In Britain, suburban expansion was promoted from the 1870s (Pollins, 1964) through legislation for cheap rail fares to encourage workers to commute, by state funding of the new physical infrastructures, by nationalisation of and invariant prices for utilities, by income tax rebates to promote owner occupation, from the 1950s by state-led construction of the New Towns, and from the 1970s by concessions on sales tax on new housing construction. In the other English-speaking countries infrastructure and tax subsidies to suburban growth were massive (Walker, 1981).

Britain differed from the other English-speaking countries, however, in having public policies which *inhibited* suburban growth: the refusal of Conservative outer boroughs to build council housing, and, post-war, the nationwide imposition of green belts and halts to urban spread 'to protect the countryside'. These constrained outward movement of the poor, a large proportion of whom have remained locked into the inner city council housing (for whites) and privately-rented slums (for immigrants) (Hall *et al.*, 1973). The New Towns did not take the poorest but favoured the 'respectable poor' and the skilled working class. To a large extent the outcome has been a *spatial separation* of the 'rough' from the 'respectable' poor – a coercive rather than incorporative approach in the terms of Chapter 1.

Some decentralisation of the poor did, however, take place. Many moved out of the East End of London, for example, reducing its density from its Edwardian peak – and helping to dilute its radical politics. In the early twentieth century some suburbs formed around a single large factory or an industrial cluster, Cadbury's Bournville in outer Birmingham and Ford's Dagenham in outer London being notable examples; some of these exhibited a successful mix of Fordist industrial discipline with the discipline of stable family life (Smith, 1989). But even then, suburbanisation largely failed to ameliorate poverty. The new manufacturing suburb was impaired by industrial decline and increased out-commuting. The classes have not mixed, even in the New Towns where housing was designed with this explicit aim (Reade, 1987: 136). In consequence, poor neighbourhoods and whole localities have emerged in suburbs and New Towns – a phenomenon found not only in Britain (CES, 1985) but in continental Western Europe (Wacquant, 1996) and, especially, the USA and Australia (Mee and Dowling, 2000). Low-density housing and greenery do not overcome poverty; *this* environmental determinism, at least, has been a failure.

2.6 THE NEIGHBOURHOOD

We saw in Chapter 1 that a key theme of poverty policy since the early modern period has been that there has been a problematic 'culture of poverty' characterising 'an underclass'. Since the industrial revolution this culture has typically been seen as reproduced at the *neighbourhood* level: the slum, the thieves' quarter, the 'dreadful enclosure', the wine alley, or, more recently, the inner city and the 'sink estate' are seen as the crucibles of deviant behaviour. Individual households may be deviant and may reproduce deviancy in their children (section 3), but this is seen as mostly because of local influences. Since class has spatially divided the city, and since the poor have had few social interactions *beyond* the city, the poor neighbourhood is the obvious culprit for deviant socialisation. Intervention to deal with poverty has accordingly been strongly targeted on these neighbourhoods.

The 'worst neighbourhoods' have been stigmatised in dominant ideology as dark and dangerous, and places the respectable should avoid. In the nineteenth century they were seen as potentially contagious: the typhoid germs lurking in the water supply were analogous to the moral contagion hidden in the recesses of the city. Indeed, the *place itself* has often been pictured as the problem, in another variant of spatial determinism (Hebbert, 1980). For the inhabitants themselves things have been, of course, more complex: the neighbourhood may be seen as home, as support network, or as a prison.

The problematic of the deviant neighbourhood emerged from both the growth *and* internal differentiation of the cities. From the sixteenth century moralists contrasted the stability, cohesion, orderliness and sense of community in the countryside with the unruly and immoral society of the city, or at least its lower orders (Williams, 1975; Hebbert, 1980), a theme elaborated in the nineteenth century in classical sociology. Clear spatial differentiation by class began in the largest towns during the eighteenth century, and was incorporated from the first into the new industrial cities. During the nineteenth century the anxieties of higher classes therefore became focused not simply on the city but on its working-class neighbourhoods. Over the nineteenth century the spatial separation between the rough and respectable working class deepened, just as intervention into working-class living conditions was beginning. Both anxiety and intervention were accordingly focused on the *poorer* working-class neighbourhoods (Stedman Jones, 1971), a preoccupation of poverty policy ever since.

Figure 2.1 Easington, County Durham, 1984. Repression of a locality during the miners' strike (source: Keith Pattison).

A *series* of distinct contrasts at different spatial scales has thus been central to the development of this ideology.

While informed by this common stance, interventions into poor neighbourhoods have ranged from the coercive to the incorporationist (section 1.3). Let us consider these in turn.

Containing the poor neighbourhood: repression

Repressive policies towards poor neighbourhoods range from isolation and abandonment to policing to eviction and eradication. They are most used when there is no realistic prospect of improving the neighbourhood's condition due to national or local economic crisis. **Isolation** of the neighbourhood *as a policy* is motivated by a wish to prevent the poor from contaminating the culture of the respectable working class, and to reduce state expenditures. Isolation is largely *achieved* by the markets in housing and private sector consumer services coupled with popular revulsion against the poor (sections 6.6 and 7.2). But this can be deepened by state policy, depending on the spatial structures of funding. In the USA, for example, it has long been the case that there are big differences in funding of public services between rich and poor local authority areas, and between neighbourhoods within them (Fainstein, 1994). Where a local economy collapses public investments and services often decline too, in order not to 'waste' money on people whose labour power is unusable (Davis, 1992: 304–7); the state thereby signals an official abandonment of the area. In the present period, neoliberalism has tended to worsen the public services for the poor, functioning to further stigmatise them (section 5.4).

If jobs and good public services are not to be provided, discontent may need to be held down by **policing and 'security'** (Neocleous, 2000). Altering the built environment may help to do this. New housing may be designed to be readily policed. Major roads can serve the same purpose. Haussmann's boulevards in Paris, which dissected the medieval street pattern following the difficulties of the army in penetrating the plebeian neighbourhoods in the 1848 revolution, provided the model for planning in other European cities. Following the 1992 riots in south-central Los Angeles, during which missiles were hurled onto the freeways, a freeway raised on stilts was built across the area to enable residents in adjacent neighbourhoods to escape future riots. Similarly, the rapid transit system serving the central business district (CBD) of Detroit was put underground through the poor inner neighbourhoods and had no stations in them.

The direct policing of poor neighbourhoods has also often been heavy. In nineteenth-century London army barracks ringed the poorest neighbourhoods. Present-day black neighbourhoods in Britain (Sivanandan, 1990), as well as France and the USA (Davis, 1992; Wacquant, 1996), have been policed arbitrarily and violently, against the wishes of the majority of the population. Often policing is directed towards preventing people using the neighbourhood's streets for social – potentially political – interaction, sometimes using curfews;

the 'sus' law was used in this way in Britain in the 1980s, and the present Labour government has introduced laws to facilitate this. In a related, though inverted, geography, police harassment of Romanies and New Age travellers protects respectable neighbourhoods from their encampments. Policing sometimes extends to attacks on political organisations in poor neighbourhoods: police infiltration of Communist Party branches was common in many countries during the twentieth century; in the USA in the 1960s the Black Panthers' community organising of the poor was destroyed by armed force. When the state turns repressive it would rather see social disintegration in poor areas than political organisation. All these forms of policing resemble those used against colonial populations, and the imperial states have often transferred their acquired expertise from the one context to the other (cf. section 2.2). Thus in the early 1970s there was much discussion about the need to bring the 'lessons' of policing Northern Ireland to bear on disorderly localities in Britain.

If all else fails, the poor may simply be **evicted** from a neighbourhood and **dispersed**. The worst 'rookeries' of nineteenth-century London were forcibly cleared and demolished on social-political grounds. Commercial property development around central business districts has often functioned to evict and disperse the poor, as has expansion of industry and universities into residential neighbourhoods. In the case of the building of the railway termini in nineteenth-century London and many provincial cities, the sites were chosen partly for this aim; the same was true of the recently constructed new commercial centre of Bunker Hill in Los Angeles. In the present period, many cities are encouraging gentrification of inner neighbourhoods to disperse the poor (as well as to foster a new city culture: section 10.4). Even the social democratic slum clearances of the post-war period, with their rebuilding of working-class (social) housing in situ, sometimes had a hidden agenda of dispersing the 'underclass' and weakening neighbourhood culture (Palmer, 1972: 27) and typically had that effect: people were 'temporarily' moved for rebuilding, but most never came back. Through these varied means, the 'worst areas' are eradicated.

All these forms of repression deliver the same message: conform and work, or be punished.

Improving the poor neighbourhood: incorporation

In contrast to the repressive approach, incorporation has attempted to improve the poor neighbourhood. Provision of education, health services, social work and (again) better housing has attempted to improve the neighbourhood's way of life and healthiness, partly aimed at better employability. Intervention has mobilised peer group pressure through religion and community building. Sometimes rewards for changes in behaviour have been used, for example by making the provision of social housing conditional on good conduct. Environmental determinists have tried to change culture by changing the neighbourhood's built environment.

The mix of these different forms of intervention has changed over time. In the late eighteenth and early nineteenth century moral exhortation by the churches was the key intervention, directed at instilling the work ethic, self-improvement and birth control – 'marrying late, breeding late, and using a savings bank' (Hebbert, 1980: 7). Methodism was particularly successful in this because of its attention to local community. From the mid-nineteenth century social work, schooling and health visitors took over these tasks (though usually still strongly marked by religion). From the late nineteenth century through to the 1960s intervention in poor neighbourhoods became strongly centred on **housing and its layout**, reflecting the prominence of environmental determinism in British social thought both left and right: an improved environment would change the behaviour of the working class. The slums were dangerous because social evils could flourish where they were not exposed to the healthy gaze of 'society'. By replacing the courts and alleys with orderly housing, open to light and air, the poor could be regulated by observation and influence of better educated neighbours; adopting the principles of the panopticon, the poor would be improved by becoming visible to their betters. The environmental approach reached its apogee after the Second World War, with an enormous programme of slum clearance and rebuilding (as well as suburban expansion: section 2.5). Combined with expanded public services, new public housing was laid out in such a way as to promote social interaction, a sense of community, and ultimately civic mindedness. New and renewed working-class communities were to be physical symbols of the social democratic 'classless society'.

The 1960s and 1970s, however, demonstrated the limitations of physical renewal, as poverty was 'rediscovered' and unemployment swept through the areas that had been renewed. There was a return to nineteenth-century prescriptions for poor neighbourhoods based on **cultural and social change**. Neoliberalism, using coercive measures, promoted work, enterprise and self-reliance for the poor individual and household, ignoring the neighbourhood scale. From the 1990s conservative interventionism sought to temper, though not over-ride, this regime with elements of incorporation. These were strongly focused on the neighbourhood, in the belief that poverty is created, or at least perpetuated, primarily by local cultures of poverty rather than individual failings. CI has thus promoted revived community organisation, locally-based social enterprises to provide services and intermediate level jobs, and greater political participation by the poor. The neighbourhood has once more become the centre of poverty policy through community regeneration.

A common trait of neighbourhood policies whether coercive or incorporationist is their tendency to **spatial determinism**. The identification of poverty with the 'worst' neighbourhoods implies, counter-factually, that it is not significant elsewhere. It implies that poverty is created for the most part *within* the neighbourhood, either by the built environment or by some *sui generis* local culture. This prevents the policymaker having to consider deeper, and more politically difficult, explanations of poverty involving higher spatial scales (cf. section 1.2). The focus on the slum or the sink estate as physical and moral problems has allowed reformers to avoid considering what Booth called 'the quagmire underlying the social structure' (quoted in Yelling, 1986: 2). It legitimates quick fixes focused on 'obvious' causes of poverty such as the environment, crime or lax morals. The state can then be 'seen to be doing something'. Most importantly, the territorial focus deals with a *real* threat for the elite, namely well-organised collective resistance by poor neighbourhoods (section 12.2). The lurid vision of disorder and social decay in these areas expresses, above all, this fear.

2.7 CONCLUSION: CONTORTIONS OF SPACE IN THE MANAGEMENT OF POVERTY

We see, then, that both problematising space and the use of space have been integral to poverty policy. There has been considerable continuity over the history of capitalism in these geographies. First, policy and ideology have operated at every **spatial scale** from the globe to the individual. These scales are strongly connected to each other by policy: the nation is mobilised to compete internationally, poor regions and localities are addressed in order to improve national economic and social reproduction, suburban expansion in the city-region resolves problems of poor neighbourhoods, policies for poor neighbourhoods remove a sore on the body of the city and nation, and so on. We shall see in Part II that these scalar connections are, in part, a result of the economic and social *construction* of poverty and exclusion at multiple, interacting scales.

We noted at the beginning of this chapter that policy and ideology deploy space in two distinct modes, as **territory** and as **flows**, and we have seen many examples of these at different scales. Both modes have characteristic advantages for capital, but they also contain systematic dilemmas and tensions. Let us briefly summarise these:

1   The contrast between 'poor' and 'normal' places

- This contrast can be made for territories of very different scales, from nations to neighbourhoods. These then tend to produce different stereotypes of the poor: these can be *British* workers in general (pricing themselves out of the market), those in de-industrialised *regions* (Thatcher's 'enemy within'), pockets of deviants in problem *neighbourhoods* (the police's 'criminals or cowards'), or particular *minority-ethnic groups* (almost everybody's 'other'). The variety of discursive and policy scales contributes to the ambiguity of the concept of the 'underclass' (see section 1.5).
- A given area can appear as normal or deviant according to the scale of the discourse and its political strategy. 'Defending Britain's interests abroad', for example, constitutes all Britons, including the poor, as 'one

of us' *in contrast to* foreigners. But discourses which picture Britain's 'cultural decline' as embodied in an expanding underclass divide Britons between the normal and the deviant, as do discourses of inner-city poverty. Different class projects thus produce different socio-spatial divides and different borderlines of inclusion and exclusion.

- A given policy may be applied at different spatial scales. Stimulating 'enterprise', for example, may be applied at the British level (because Britain is dominated by large corporations), the regional level (because old industrial regions lack enterprise) or the neighbourhood level (because they have a culture of poverty). The comparison of 'good and bad' areas can be performed at each spatial scale. The rational core here is that 'lack of enterprise' is a *social relation*, which is ubiquitous and may operate at any spatial scale (Gough, 1991).

- The comparison of areas implies that poor nations, regions and localities are host to dysfunctional social relations *in contrast to* healthy ones in non-poor areas. The spatial representation of the difference between poor and non-poor literally distances the latter from the former. Places with stable forms of exploitation and oppression are constructed as good *just because* they are not poor. The spatial contrast legitimates both changing the poor *and* maintaining the status quo for the majority.

- The spatial contrasts elide place and person. Is it the area or the people that are poor? This ambiguity allows poverty to be constructed as *inherent in the place*, an unmediated product of the territory's environment or history, rather than *its place within* power relations and the systematic uneven development they produce. It legitimates solutions confined to that scale (cf. section 1.2).

- A consequent ambiguity is whether policies are 'for the place' or for the people in it, and if the latter, which people? The notion of a problem *area*, whether a country, region, locality or neighbourhood, obscures class, 'racial', gender and other differences in interest *within* the area.

2   Regulating flows of capital and people between areas as solutions to poverty

- There are enormous difficulties in both stimulating and preventing flows. In capitalism, firms command their capital and its investment, while most workers control the disposal of their person and their labour power. Preventing capital and labour flows may contravene fundamental rights and be highly politically contentious. Stimulating flows may require costly incentives and support. A corollary of these difficulties is the dilemma of whether to focus on the flow of capital or the flow of labour. Should capital be encouraged to remain in poor areas and flow to the poor; or should the poor be encouraged to move to capital and to remain where investment is strong?

- It is often unclear whether the regulation of flows is to benefit the poor or business. Is bringing inward investment to a poor region aimed at providing employment for the poor or supplying low-paid labour for firms (Massey, 1979; Dunford and Perrons, 1983)? Is promotion of immobile forms of capital such as cooperatives motivated by stabilising employment for the poor or encouraging their self-exploitation (Eisenschitz and Gough, 1993: 161, 208)? Is facilitating movement of the poor to growth regions aimed at getting them employment or solving labour shortages?

These ambiguities and elisions of policy focused on territories and flows can be politically useful in legitimating interventions into poverty and in enabling 'easy' policy solutions to be implemented. But the ambiguities also point to real political *dilemmas* for capital and the state in their spatial strategies (cf. Harvey, 1982; 1989; K. Cox, 1993; 1998). To what extent should policy attempt to impose spatialised discipline, territorially-based sacrifice and spatial exclusion on the poor? To what extent should it improve the material welfare of the poor, encourage their community and political involvement in territory, and spatially integrate them? We can now see that the contrast between coercive and incorporationist strategies, between liberal and social democratic approaches, which we traced in Chapter 1 carries through to the use of space. Indeed, the different political strategies use space in distinct ways, and the latter are essential to the contrast between the approaches. We shall further examine this internal relation of poverty strategies and geography in Part III.

3

PATTERNS OF SOCIAL
EXCLUSION

3.1 WHAT ARE POVERTY AND
SOCIAL EXCLUSION?

In this chapter we offer a snapshot of poverty and social exclusion in contemporary Britain. The purpose is to show that deprivation, however defined,
is severe, thus justifying the analysis of the rest of the book. We also wish to
describe some of the particular *aspects* of deprivation with which low income
is associated as results and, to some extent, causes. Further, we describe how
poverty affects different *social groups* differently, in particular, how social
oppression renders people more vulnerable to poverty. Making these distinctions
helps to locate poverty more concretely in social life; but explanations are
developed in Part II.

In the Introduction we recounted the shift in the dominant discourse from
'poverty' to 'social exclusion', and discussed some pros and cons of these terms.
This discussion suggested that all terms for deprivation are politically loaded.
What, then, do *we* regard as useful? '**Absolute poverty**' refers to insufficient
resources for physiological reproduction: food, shelter, a healthy physical
environment, and care of the body. The concept is popular because it can be
asserted that absolute poverty has been overcome in the developed countries.
But the poor have a shorter life expectancy and worse health than the better
off. Indeed, the *majority* of the population in the developed countries have their
health and life span negatively affected by such social conditions as diet, air,
water, work and stress. Physiological reproduction is not fully achieved by
present-day society, and this shortfall is especially large for the poor; this can
then severely affect participation in work and social life. Absolute poverty is still
with us.

Because absolute poverty in developed countries is *less* than it was, and less
than it is in developing countries, poverty studies tend to use the concept of
'**relative poverty**'. This is an income less than some proportion of the national
average; the *distribution* of income, rather than its absolute level, becomes the
key political consideration. This, too, is a politically useful measure of poverty.
People compare their standard of living with others in the same national space,
and feel that they have been demeaned or are failures if they have much lower

Figure 3.1 Street scene in Newcastle upon Tyne, 1962. Poverty, but strong sociality: a kind of inclusion? (source: Newcastle Chronicle and Journal).

income than the average. These feelings appeal to notions of *justice* – that all who contribute to a society deserve a reasonably equal reward – and to notions of *respect* – that giving little to some people insults them (Baker, 1987).

In recent years, however, poverty has come to be seen as relative to *the demands society makes on the individual*. This is embodied in the UN's (Gordon, 2000: 50ff) concept of **'overall poverty'** and Sen's (1983) notion of **'hybrid poverty'**. Its roots go back to Rowntree's argument from his second survey of York, that poverty should be defined by the minimum resources needed to participate in the main social practices of the society. It has two facets. First, it focuses on **concrete patterns of consumption** of the poor enabled by the society, not just their incomes. Particular items of consumption may be more or less accessible in the given society. This is important where items such as housing consume a high proportion of income; conversely, some poor groups may be kept out of poverty because they live in cheap public housing. Similarly, people on moderate incomes in the USA may be unable to afford adequate health care and hence are poor in (at least) this respect.

Second, 'overall poverty' considers **people's ability to participate in normal social interactions** – social, political, cultural and economic life. The long-term increase in the average standard of living in the developed countries has not merely furnished people with an increasing number of luxuries which poorer people can easily do without. Rather, increased consumption of commodities and public services *changes the patterns of social activity*, and hence to a considerable degree become *necessary* for participation in essential activities. Numeracy and literacy have become necessary for most jobs; many people cannot access jobs without a car; social intercourse is curtailed if one does not watch TV or have a mobile phone. Sen argues that while there are universal human needs, societies vary in the particular expression of these needs and what individuals materially require. Thus education and learning are universal needs, but their content varies between societies. It is thus possible to define a minimum *social consumption norm* which is compulsory for *social participation* in any given society at a particular time.

In our view this is the best way of conceiving of poverty. It bridges the absolute and relative concepts of poverty and renders them obsolete (Gordon, 2000: 51). It also makes the conception of poverty more complex:

- People who live, or seek to live, ***different kinds of life*** require different minimum resources to do so. The poverty threshold differs between social groups, stages in life, household forms, and patterns of work and social life.
- ***Particular combinations of resources*** have unique properties. A single parent may be allocated social housing, but its location away from her family and social network cuts off informal childcare and makes it impossible to take up waged work. Thus, even improvements in one dimension can worsen others.
- Certain resources give benefits which, in contemporary society, are inherently ***comparative or 'positional'***, so that their acquisition does not necessarily give a *useful* resource. Thus the improved access to jobs resulting from better education is continuously negated by the inflation of qualifications required for given jobs. Owner occupation, which once gave status and financial stability, now often confers neither: half of those in poverty are owner-occupiers (Burrows, 2003). Again, social context needs to be considered.
- Disadvantage in a particular dimension such as education or health, or lack of a particular resource such as adequate housing, may ***cause or perpetuate low income***. More widely, oppressive social relations create conditions which aggravate poverty; the poor are, for instance, more likely to become disabled than those with higher incomes (Burchardt, 2003). Recent definitions of poverty thus tend to include both its concrete effects *and* some of its **causes**.

The last point indicates an element that many definitions of deprivation fail to take into account: that basic consumption norms should be met through **non-oppressive social relations**. If adequate consumption is achieved through oppressive, hence damaging, interactions, then poverty has not been overcome. Thus, a job may pay an adequate wage but destroy the autonomy, health and self-respect of the worker. Similarly, welfare services have often disempowered their clients (see section 5.3). These resources are therefore inadequate solutions to poverty.

These considerations lead from poverty as low income to poverty as a barrier to social participation, and hence to the concept of '**social exclusion**'. As we noted in the Introduction, social exclusion *may* be interpreted broadly and radically. It can suggest that lack of resources tends to exclude the person from normal types of social interaction. It can suggest that social relations affect people's income. Social exclusion can also take us beyond the purely economic, to suggest that gender, racism, homophobia and the social construction of disability and age are, as well as causes of low income, important forms of exclusion *by virtue of* inhibiting social participation. Indeed, social oppressions limit social inclusion even for people who are not on low income.

However, the interpretation of 'social exclusion' by the Labour government is narrow and conservative. *Social participation* means obtaining a job, however poor, and participation in normal political activity rather than rebellion; other forms of inclusion are seen merely as means to these ends. *Social relations* are seen as generating poverty through individuals' lack of 'social capital'; the social relations that produce shortage of employment, poor jobs, low levels of benefits and discrimination are ignored. The notion of *social oppressions* is also weak in Labour's thinking. 'Social exclusion' is a highly ambiguous concept.

A further weakness of Labour's conception is that it ignores the ways in which some of the traits of poverty **extend into the mainstream**. Unaffordable housing, the erosion of private and state pensions, insecure or oppressive jobs, and racist or sexist violence are also experienced by people who are not poor. This problematises *any* demarcation between 'the poor' and the 'non-poor', and at least renders the boundary fuzzy.

Our view, then, is that the conception of poverty needs to focus on **oppressive and exploitative social relations**: how they create low income and inadequate consumption; how they affect particular experiences of deprivation; how lack of consumption items affects social participation; how oppressive social relations are themselves a part of deprivation; and how aspects of poverty therefore extend into the mainstream. This does not so much define poverty as define our *approach* to analysing and changing it which we develop in the rest of the book.

This discussion suggests that different notions of poverty connect with different political strategies, though in very complex ways (see Table 3.1); this is explored further in Part III. Nevertheless, all the different conceptualisations can contribute to our understanding. Absolute poverty points to societal processes which harm and destroy bodies. Relative poverty highlights the injustice and lack of respect inherent in unequal income distribution. Hybrid or overall poverty focuses on the disruption of patterns of social activity by lack of material resources. Social exclusion can be understood to suggest that low income is caused by oppressive social relations. In the rest of the book we shall therefore use 'poverty', 'social exclusion' and 'deprivation' loosely, except where we are discussing their more precise usage by particular political strategies.

If the traits of poverty and risk of falling into it reach into the general population, and if mainstream social relations are important in generating poverty, then our discussion cannot only concern '**the poor**'. We will, in fact,

Table 3.1 Different concepts of deprivation

Concept of deprivation	Measure	Politically-significant effects	Possible policies
Absolute poverty	Income so low as to substantially harm physiology	High mortality, ill health, inability to work	Policies to increase income; subsidised basic resources – via state *or* change in economic structure
Relative poverty	Income distribution; ratios of incomes	Social inequality and injustice; disrespect	Income redistribution – via state *or* change in economic structure
Hybrid or overall poverty; social exclusion 1	Low income; lack of particular resources	Material deprivation, as above; constraints on social and economic participation	Policies to increase income, as above; subsidised basic resources; welfare services
Social exclusion 2	Social oppressions: gender, racism, age discrimination, etc.	For oppressed groups: constraints on social participation; material deprivation	Combating discrimination; *or* combating roots of social oppression
Social exclusion 3	Lack of cultural and social capital	Material deprivation; constraints on social, economic and political participation – these may *cause* low income	Improving social, community and political participation of the poor
Deprivation constructed by the whole society	Sharp exploitation; social oppression; disempowerment; alienation	As above	Increasing the power of the poor; changing the fundamental social relations of the society

often discuss deprivation in 'the working class', in the usual sense of manual and low-status service workers and their families. Much of our discussion will be of 'the working class' in the Marxist sense of those who depend on wages directly or indirectly, including most of 'the middle class'.

In the rest of the chapter we describe some aspects of deprivation for which data are available. The aim is to make vivid some of the phenomena that are to be explained in Part II. For more comprehensive surveys see Gordon and Pantazis (1997), Gordon *et al.* (2000), Hills *et al.* (2002), Palmer *et al.* (2004) and Flaherty *et al.* (2004), and on the sub-national geography Green (1994), Philo (1995) and Hudson and Williams (1995).

3.2 INCOMES

We may start with the British government's main measure of poverty, those living in households with income below 60 per cent of the median – a measure of **relative poverty**. On this criterion, in 2001/2 22 per cent (12.5 m people) were in poverty, including a third (4.3 m) of all children. Relative poverty fell during the 1960s, increased again in the 1970s, and grew dramatically in the 1980s: whereas in 1979 13 per cent of the population was poor, by 1992 this figure was around a quarter where it has remained ever since. Changes in **income distribution** also portray increasing inequality. Between 1979 and 1999 the share of the bottom decile fell from 4 per cent to 1.9 per cent, while that of the richest increased from 21 per cent to 29 per cent (Levitas, 2001); the latter has 15 times the income of the former. But this is not just a matter of the extremes: the bottom third of the population shared 16 per cent of national income in 1979 but only 11 per cent in 2000, while the top third increased their share from 48 per cent to 56 per cent. In absolute terms, **real incomes** for the poorest decile *fell* by 17 per cent between 1979 and 1991, while the incomes of the top decile rose by nearly two thirds. Trends in the USA have been even worse; for example median income for black families with children headed by people under 30 *halved* during 1973–90 (Sklar, 1995b: 19).

The post-1997 Labour government has redistributed income to some of the poor through tax credits, principally to working families with children and to pensioners. The number of children in poverty has fallen from 24 per cent to 16 per cent between 1997 and 2004, while the number of poor pensioners has halved (Sutherland *et al.*, 2003). But employed people without children and the unemployed have not gained. During 1997–2001 the poorest tenth saw a 13 per cent increase in income, though one quarter of them suffered a loss of income (Levitas, 2001). The overall distribution of income has worsened because fiscal redistribution has not been at the expense of the well-off and because inequality in wages and unearned incomes continues to increase. Thus between 1997 and 2002 real incomes before housing costs increased 10 per cent for the poorest decile but 17 per cent for the average (Flaherty *et al.*, 2004: 45).

An aspect of 'overall' poverty is measured by constructing **a consensus over what constitutes essential subsistence**. This produces similar results. On this definition, 14.5 million people or 26 per cent of the population were in poverty in 1999, defined as lacking two necessities, including such things as a damp-free home and being able to visit friends and family in hospital (Flaherty *et al.*, 2004: 31). Four million people are inadequately fed, six and half million go without essential clothing, and a further million cannot afford social activities. Seventeen per cent of a representative sample of the population perceived their own incomes to be below the UN-defined poverty line (Gordon, 2000: 55). These shockingly large figures are confirmed by descriptions of life in Britain as a struggle for survival (Danziger, 1997).

There is also large **movement of individuals** across the official boundary of poverty. Over the 12 years from 1991, half the population spent a least one year in a poor household, 20 per cent spent at least five years and 1 per cent spent the whole period (National Statistics, 2004). The majority of the poor move in and out of poverty, but with little *substantial* upward or downward mobility. Around a third of the population spend a large part of their lives in or near poverty.

3.3 EMPLOYMENT AND THE STATE'S REACTION TO IT

The cause of low incomes depends upon the position of three groups of adults: those in low wage employment; the inactive who are seeking work; and adults not seeking employment due to personal incapacity or retirement. The majority of the latter two groups are dependent wholly or partly on state benefits or pensions. Children's poverty derives from being in households with these adults. Poverty is then crucially dependent on two features of the labour market, wages and unemployment. State benefits for those of working age are not independent of these conditions, since they are set below the lowest wage.

All these causes of low income have worsened in the last three decades (explanation, and more detail, are given in Chapters 4 and 5). First, **the quality of employment** has deteriorated. Both part-time and casual employment have increased: between 1975 and 1993 there was a fall in the proportion of the population holding full-time permanent jobs from 55 per cent to 35 per cent (Mohan, 1999: 93). Differentials in hourly wage rates have increased (Nickell, 2004). According to Hutton (1995) only 40 per cent of wage earners have stable employment capable of supporting a family at a reasonable living standard. Moreover, it is difficult for the poorly paid to improve their wages: the unskilled are three times less likely to receive work-related training than those with qualifications (Howarth *et al.*, 1999: 57). These trends make it important for two-adult households to have two full-time wages to keep out of poverty.

Second, **unemployment and economic inactivity** have been major causes of poverty in the last 30 years. Data adjusted for the many changes in defining unemployment show unemployment at 3.4 per cent in 1973, climbing steadily to 12 per cent in 1984 and dropping to 4.6 per cent in 2004 (Lindsay, 2005). The last is significant since the number in employment is at its all-time high of 28.5 m. But these aggregate figures hide the extent and impact of unemployment. Unemployment affects ever more groups: in a four-year period during the early 1990s 10.6 m people experienced a spell of unemployment. Moreover, the rate of economic inactivity of those of working age is currently 21.5 per cent and has changed little since the 1970s. Of the 8 million people in that group, over 2 m say that they want to work, and it is likely that many more would want to if appropriate jobs were available. Economic inactivity has risen particularly sharply for the over-50s, from 7 per cent in the mid-1970s to 23 per cent in 2000; in contrast, there are more young people counted as 'economically inactive' in make-work schemes or in full-time education – the majority on very low incomes. Good quality work is now concentrated in the 25–50 age group (Beatty and Fothergill, 1998: 140). Many of the economically inactive are on incapacity benefits; their number has increased over 25 years to 2.5 million; in the 55–64 age group nearly three quarters of economically inactive men say that sickness or disability prevents them working, over double the rate of the 1970s (Nickell, 2004). As a result of these trends, more than one in six households with two adults of working age have no-one in employment, a far cry from the two-wage norm.

Third, **demographics** have tended to increase the proportion of the population in poverty. The birth rate has fallen faster for the better-off than for the poor. The proportion of the population in retirement has increased, and the retired are more likely to be poor than those of working age (section 3.5 below).

Fourth, nearly all **state benefits and pensions** have declined relative to average earnings since the 1970s. Between 1977 and 1990 unemployment benefit declined from 48 per cent to 32 per cent of average net earnings, and income support from 36 per cent to 20 per cent (Dean and Taylor-Gooby, 1992: 59; Becker, 2003). Between 1981 and 1997 the state pension fell from 32 per cent to 16 per cent of average male earnings. These cuts have had a major impact on poverty: since 1979 there has been a large increase in poor households headed by claimants.

Contemporary poverty, then, has its immediate causes in labour markets and state benefits policy. As these change, so does the composition of the poor. In the 1960s, buoyant labour markets restricted poverty to those not looking for work – half the poor were pensioners. Since then, not only has the number of the poor increased but the unemployed and low paid constitute a higher proportion of them. Between 1979 and 2002, among the bottom decile of incomes pensioner couples fell from 20 per cent to 6 per cent, while working-age single people without children rose from a tenth to a quarter, and the unemployed and inactive rose from a third to 45 per cent (Flaherty et al., 2004: 44). These changes suggest that we should not assume that any particular social group is essentially poor, but rather investigate the economic, social and state *relations and processes* which put groups into poverty in historically- and spatially-specific ways (see Part II).

3.4 ASPECTS OF DEPRIVATION AND THEIR RELATION TO INCOME

According to the 'hybrid' or 'overall' approach to poverty, aspects of deprivation such as housing, nutrition or education cannot be read off from low income although they correlate to a considerable extent. The causation can be both ways: poor health, education and mobility are a result of low income but in turn harm participation in waged work. Let us consider some of these aspects of deprivation and their correlation with low income. Notice that some of these are true of poor *neighbourhoods*, some of poor *households*, others of poor *individuals*. Space in the senses of *enclosure* and *distance* is often involved. (Explanation, and more detail, is given in Chapter 6.)

The poor tend to live in the worst **housing** in a given locality: even though social housing is in many instances of good standard (Folwell, 1999), around three quarters of the housing inhabited by the poor is unfit. Since deregulation in the 1980s, rents in both private and public sectors have soared, and the social housing stock has declined. Over the same period, high inflation in house prices has put owner occupation, except in the very worst areas of the poorest regions, out of the reach of the poor (Thomas and Dorling, 2004). The consequence has been rapidly increasing homelessness: around 500,000 people currently live in atrocious temporary accommodation, on friends' floors, or on the streets). **Keeping warm** required over 3 m poor households to spend more than 10 per cent of their income in 2003 (Flaherty *et al.*, 2004: 113). The poor not only have higher fuel needs because they spend more time at home but are also more likely to live in homes that are expensive to heat.

The poor spend a much higher than average proportion of their income on **food** (Dowler and Turner, 2001). Hunger is still common and poor people's diets are inferior: official guidelines on healthy eating show that a family on income support would have to spend nearly 40 per cent of its income on food. Poor housing, heating and diet are obvious causes of the **bad physical health and high mortality** rates of the poor. Twenty-five to forty thousand poor pensioners annually die of cold (Carley and Kirk, 1998: 9). Poor neighbour-hoods have higher than average exposure to exhaust fumes and industrial pollution. The poor are also more susceptible to poor environments: in the Chicago heat wave of 1995 700 deaths were caused by poor housing and the privatisation of public health (Klinenberg, 1999). Accident rates are higher among poor children, and among low paid and casualised workers. The high levels of stress among the poor, and the higher cigarette consumption it leads to, damage physical health. Casual workers lack sick pay and job security which forces them to work through illness, exacerbating the initial problem. Those without work tend to have a 20 per cent higher death rate than the employed. Consequently, the incidence of most **diseases** increases as income falls (Widgery, 1993), often dramatically so. In one survey of caries, lower-class 3-year-olds had nine times its incidence compared to the higher social classes (Nadanovsky, 1993). The **mortality** rate is higher for the poor at all ages, and has worsened in the last 30 years as poverty has increased (Rahman *et al.*, 2001: 53).

Public services such as schools, health and social services, nurseries, and care of the elderly and disabled tend to achieve the worst outcomes for the poorest people, particularly but not only in poor localities (Carrier and Kendall, 1998; Kumar, 1997). Pass rates in school exams in poor neighbourhoods, for example, are less than half the national average (Lupton and Power, 2002: 124).

At a time when wealthier groups are becoming 'superincluded' by **financial institutions**, the poor are systematically denied access (Leyshon and Thrift, 1996). Yet they need financial services more than most: insecure incomes, inability to purchase durables from current income, and coping with the unexpected require credit. They usually cannot draw on savings for these needs: one third of all families have no savings. Half the poorest fifth of the population cannot afford house contents insurance yet they are three times more likely than the average to be burgled. The gateway to the credit that would remedy this, is through banking; yet 1.5 m adults lack basic banking facilities while another 4.4 m exist on the margins of financial exclusion (Rahman *et al.*, 2001). The poor therefore often end up borrowing from loan companies at extortionate interest rates and consequently never get out of debt. **Other private services** consumed by the poor are often inferior in quality and they tend to pay more for them. Retailers and banks increasingly concentrate on better-off neighbourhoods (Speak and Graham, 2000). Lacking bank accounts or due to debt, the poor are often charged the highest tariffs by utilities.

The **communication** of the poor is restricted. Few poor households have cars, and for those who require one – for example in rural areas – the expense can *make them* poor. Bus and train fares as a proportion of their income are very high and limit access to important economic and social activities (Palmer *et al.*, 2004). Holidays and visits at a distance are usually unaffordable. The social interactions of the poor are thus spatially limited, and they are often confined to their home and immediate neighbourhood, something that the stigma of poverty reinforces.

People living in poor neighbourhoods suffer from higher **crime** rates – burglary, street robbery, physical assault, murder, and vandalism; policing is weaker in these areas. They are also more likely to experience **serious nuisance** from their neighbours such as persistent noise, intimidation and threatening behaviour, which can cause severe stress.

Relationships of partners, with or without children, tend to be more unstable for the poor than the better-off; in the USA they are twice as likely to break up (Sklar, 1995b: 13). Poor and cramped accommodation, ill health and disability, conflicts over money, and depression frequently put enormous pressures on poor households, while the absence of wealth removes a reason to remain together. The poor in depressed regions are under particular pressure to migrate to get a job, often splitting up the household.

It is commonly argued that the poor have fewer **social interactions** and that poor neighbourhoods consequently suffer from a lack of 'social capital'. The evidence on this point is complex. Burchardt *et al.* (2002) found that 28 per cent of people in the bottom quintile of income are socially isolated, nearly double the number in the top quintile. Gordon *et al.* (2000: 60) argue that cost inhibits social interaction for a quarter of the entire population. For those in jobs or with major domestic and caring responsibilities, long hours of work and exhaustion tend to limit sociability. Fear of neighbours inhibits many, especially from oppressed groups, from going out. Generalised stress and low self-esteem may also limit social ties (Madandipour, 1998: 78). However, there is no evidence to suggest that there are 'underclass areas' where social interaction is breaking down (Richardson and Mumford, 2002; Shaw *et al.*, 1999: 206). While there exist 'estate cultures' of anti-social behaviour and crime, this is of a minority (Page, 2000). The poor generally have strong community ties, reciprocity and social organisation, and are closely connected to their own locality, however poor it may be (Cattell and Evans, 1999; Johnston *et al.*, 2000). Indeed, these networks are a key part of coping with poverty. This question is examined further in Chapter 6.

As should now be evident, the aspects of poverty so far considered often *compound* one another, and tend to perpetuate low income in vicious circles. All the processes together tend to transmit poverty and deprivation **from one generation to the next**. Childhood poverty is the most important variable in explanation of life chances: it not only correlates with subsequent low wages and high risk of unemployment but also criminality, poor mental health, and, for women, lone parenthood. It also correlates with poor cognitive abilities in adults on some measures, suggesting negative impacts of diet, physical and mental health, and socialisation in poor households. Poor children's subsequent labour market position is strongly affected by their lower educational attainment (Glennerster, 2000). Unsurprisingly, then, the *longer* a family is poor the more likely is transmission of poverty to its next generation.

The most difficult thing to measure is also the most important: the **experience** of being poor. Suggestive indicators are that clinical depression and other mental illnesses are higher among the poor, as are apparent suicide rates (Kreitman *et al.*, 1991). International comparisons suggest that the quality of many aspects of daily life depends not only on material factors but on people's sense of belonging. This may vary between countries. Gordon and Pantazis (1997: 14) argue that health can be better and crime lower in poorer but more cohesive societies with a more equitable distribution of income. This might explain why a voluntary sector worker returning to Britain from a lifetime of working in the Third World exclaimed that she had never seen such despair as she encountered in Glasgow. Differences between neighbourhoods and individuals in friendship networks can be crucial. Despite high rates of poverty among pensioners, their experience of life is often good even in poor areas, reflecting the strength of their social networks (Bowling *et al.*, 1997). Conversely, for a given income, women who are substantially isolated by caring work often find that isolation worsens the experience of poverty (Becker, 2003: 104). In this sense, too, social relations are crucial to exclusion.

3.5 SOCIAL OPPRESSIONS: POVERTY AND SOCIAL EXCLUSION

Broad notions of social exclusion are centrally concerned with social oppression, particularly those of 'race', gender, sexuality, disability and age. Social oppression often causes or exacerbates poverty; it excludes from certain social interactions; and it subjects the person to vilification, sometimes leading to low self-esteem. Below we sketch forms of deprivation suffered by certain socially oppressed groups. We should remember, however, that the effect of social oppression depends strongly on income: the better off are able to shield themselves from some aspects of oppression, while poverty exacerbates them. In each group, then, there are large differences of class position.

Black and minority-ethnic (BME) people

Most minority-ethnic groups are poorer than the British average; for example, 80 per cent of Pakistanis and Bangladeshis are poor (Cabinet Office, 2000: 18). Their labour market position is worse, for some communities *much* worse (SEU, 2001b: 46). BMEs are disproportionately employed in low wage, casualised or unhealthy jobs. Youth unemployment is over 50 per cent for Afro-Caribbeans, Pakistanis and Bangladeshis (SEU, 2001a: 92); immigrants' rate of economic inactivity is eight percentage points higher than for UK-born workers. Many BME people have worse housing and neighbourhood environments than the white population (SEU, 2001b). For some – most dramatically Afro-Caribbean boys – educational outcomes are poor. BMEs suffer worse health, exacerbated by practices of the NHS. Each stage of the criminal justice system – the police, lawyers, DPP, the courts, borstals and prisons – is systematically racist (Bowling and Philips, 2002); young blacks are nearly seven times more likely than whites to be in prison (Howarth *et al.*, 1999: 48).

Refugees

Asylum seekers are the most deprived social group in Britain. Their situation is inherently hard, but in recent years has been worsened by government policies. They are excluded from entitlement to most welfare benefits, from public housing and from mainstream education for their children, and are prohibited from working for the first six months from applying for asylum. They are provided with benefits at below income support level, and are increasingly being detained in special prisons (Fitzpatrick, 2003; Hayter, 2004). The government has also tried to eliminate the self-help that spatial proximity enables by dispersing asylum seekers from London to areas with low BME populations where hostility to them is greater.

Women

Women are disadvantaged compared to men by the nexus of inferior position in waged work and their greater responsibility for unpaid caring and domestic work. Women on average spend 23 per cent of their time on unpaid work, half again more than men; the difference in time spent in waged work is the reverse. The rate of participation in waged work is now almost as great for women as men, rising from 38 per cent to 45 per cent between 1971 and 2003, but their full-time earnings average 82 per cent of men's; 44 per cent of women are in part-time jobs, where their average hourly rate is only 59 per cent of full-time men's. Moreover, working-class women's jobs seldom offer promotion. Childcare consumes the majority of low-paid women's wages. Poor women do more unpaid work than better-off women, and therefore also have greater constraints on their waged work: thus a low-skilled woman with two children over her life foregoes earnings of over £250,000, whereas for a high skilled woman the figure is only £19,000 despite higher wages (Flaherty et al., 2004: 144). These disadvantages are particularly great for the 1.5 million lone mothers: half are poor, and they make up half of those living entirely on benefits. And with only 14 per cent of retired women receiving a full state pension, three times as many women as men are forced to live on means-tested benefits (Ginn and Arber, 1996).

Carers for the disabled and old – mostly women – constitute seven million people who save the state £57 bn a year (Flaherty et al., 2004: 95). Caring impacts heavily on the poor, including loss of household income: in one survey one half of carers lost an average of £5,000 a year. One in three are in debt, and one in five cut back on food. Council services to support carers are sparser in poor areas. And caring is likely to mean broken work records and lower pensions.

Distribution of total income within households is biased against women (Goode et al., 1998). Women are also disadvantaged within areas of consumption: men usually have first use of a car, and have better access to mortgages and private rented accommodation (Sexty, 1990). Women's inferior position in waged and unwaged work encourages generalised sexist attitudes in men, which result in patronising attitudes and harassment in public and private spaces. These conditions also make it harder for women than men to leave an unsatisfactory relationship.

Lesbians and gay men

Lesbians and gay men experience a form of social exclusion whose *immediate form* is ideological: a vilification of lesbian and gay sexuality as unnatural and of the people who 'carry' it as immoral, or as having inverted gender identity. These attitudes are often internalised as shame and self-hatred. Homosexuals experience hostility as isolated individuals, most acutely for the young, accounting for the high suicide rate amongst lesbian and gay teenagers, and sometimes leading to social isolation in later life. This oppression, then, is 'excluding' in a very direct sense.

The experience of homophobia also has material aspects. Youth who are out are usually severely bullied at school and likely to be thrown out of home by their parents. Lesbians in marriages often cannot afford to leave them. Lesbians and gay men often lose custody of their children when a heterosexual relationship breaks down. At work, harassment is common, sometimes resulting in being forced out of the job. In many developed countries there is overt discrimination against lesbians and gay men in housing, including exclusion from accommodation for the elderly. In public spaces people who are perceived as gay are often victims of violence. These forms of exclusion tend to be more virulent outside big cities in which anonymity provides protection and where communities are built, although this can adversely affect previous family and social networks (Cant, 1997). Homophobia can thus lead to poverty and exclusion from services. Conversely, poorer lesbians and gay men find it harder to escape the constraints of homophobia, particularly because of inability to afford independent housing.

People with disabilities

Twenty per cent of the working-age population – nearly 7 m people – have a disability, of which three-fifths have three or more impairments (Flaherty *et al.*, 2004: 89). The poor are more vulnerable to *becoming* disabled by virtue of their jobs, housing, and access to services. People with disabilities are excluded by the assumption that they are not capable of full social involvement, and by society's disabling failures to provide for their material needs, including the design of the built environment. People with long-term disabilities have a low rate of employment because of employers' assumptions and failure to provide support in the job. Thus only 48 per cent of people with disabilities have waged jobs compared with 75 per cent of others. The higher costs they incur in daily life – costs which are often not met by benefits – combined with employment exclusion, impoverishes a high proportion of the disabled *and* their carers (Becker, 2003: 104ff). Material exclusion makes living with disability more onerous, and reinforces patronising and disabling attitudes.

The elderly

Over the last century, the state pension, NHS and social services have enabled the elderly to be more independent. But these services are often inadequate. Older people are often forced out of employment by the physical intensity of work and by employers' prejudice. Pensioners' incomes reflect the inequalities of their working lives: the bottom quintile of single pensioners live on 21 per cent of average earnings while the top 20 per cent have nearly 90 per cent.

Female, older and single pensioners tend to be poorer. Between 1981 and 1997 the state pension as a percentage of average male earnings halved to 16 per cent (Sunley, 2000: 486); while means-tested benefits have been available to top it up, 1 m out of the 2.5 m eligible do not claim. While the housing costs of the retired are lower, other costs – food, heating, access to leisure activities – can be expensive. Moreover, some basic services give inferior treatment to the elderly; gerontology is notoriously of low medical status. Homes for the elderly, now almost all privately run, are inadequately funded (Laing, 2004) and regulated and often provide demeaning, cruel or dangerous care.

Children

Poverty is particularly cruel for children due to their mental, emotional and physical vulnerability and the impacts of harm on their adult lives. In the late 1990s one third of all children – 4.3 million – lived in households below half average income, and were short of one essential item such as adequate clothing, a healthy diet or social activities (Gordon *et al.*, 2000), a proportion which has tripled since 1979. The Labour government has effected an improvement with a fall of 3.4 per cent to 30.5 per cent in 1997–2001; but this was achieved by small increases in income for households just below the poverty line while the average distance from that line has increased. It will become, therefore, increasingly difficult to bring the rest out of poverty: the million of the poorest children have the lowest growth in incomes (Brewer *et al.*, 2003). The improvement has occurred only in households with wage earners, but in 2000 only 30 per cent of poor children were in working households (Piachaud and Sutherland, 2001). Poor families with children did not increase their spending on toys, children's clothes, and fresh fruit and vegetables between 1968 and 1995. Working-class children are five times more likely to die before the age of 16 than the better off, disadvantaged by low birth-weight, ill health, high accident rates, and family instability.

Teenagers and young adults

Young adults suffer particular forms of poverty. Working class 16–24-year-olds suffer low wages and unemployment rates that are often twice the average; for 16–17-year-old men, unemployment has fluctuated at between 20 and 24 per cent for the last decade, four times the average. Further education, the main form of training for working-class youth, is the worst funded, most casualised part of the national education system. Working-class youth are seven times less likely to go to university than children of professionals, and their results are worse. Motherhood often offers the most fulfilling role to poor teenage women, but it implies deep poverty. With young adult life now heavily structured around commodity consumption, poverty is a severe social disadvantage; non-commoditised, i.e. free, spaces such as youth clubs and playing fields are now increasingly rare. Youth is stressful for people of all classes; but poverty exacerbates this, as rates of hard drug use and suicide show. In these conditions, living with parents is inevitably full of tensions; many low-income youth leave home, often becoming homeless and in a distressing proportion of cases end up as sex workers.

In the USA – and increasingly in Britain – there is an undeclared war on working-class youth, as evidenced by the militarisation of schools and the criminalisation of minor offences previously handled by families, stores or youth workers (Giroux, 2004). Poor young males are convicted of crimes much more often than their better-off peers; there is a strong correlation – although lines of causation are unclear – between deprivation, unemployment and crime (Pacione, 1997). Four out of ten prisoners under 21 have a history of institutionalised care, one third are mentally ill, and a quarter were homeless or in temporary accommodation previous to their conviction.

Our use of these seven categories of oppression should not suggest that deprivation can be read off mechanically from membership of a particular social group, even for the poor in that group: as we shall see in Part II deprivation is produced through many *complex combinations* of social processes. These can have counter-intuitive outcomes: male unemployment, for instance, is now higher than female; average earnings for those migrant workers who do find work are on average higher than for indigenous workers. But the tendency is for social oppression to reinforce poverty and exclusion, and the latter in turn ramify the oppression – a further set of vicious circles. Moreover, social inequalities tend to reinforce each other: the emotional stress of a low-income lesbian forced to live with a man, the wholesale criminalisation of young black men, the frequent low-level criminal convictions of people with learning disabilities. The internal relations of class with social oppressions are explored further in section 7.4.

3.6 NEIGHBOURHOODS, LOCALITIES, REGIONS

Spatial uneven development always takes on different meanings when examined at different spatial scales; poverty is no exception. As one moves from larger to smaller scale, the *contrasts* in deprivation become larger – and the temptation to *explain* deprivation *by* space becomes greater. The smaller the area, the greater are the extremes at both ends of wealth and poverty. Indeed, we saw in Chapter 2 that the spatial scale at which poverty is perceived is bound up with political strategy. In particular, differences between regions and localities tend to be emphasised by anti-poverty strategies focused on supply of jobs, while differences between wards and neighbourhoods are highlighted by strategies addressing 'the culture of poverty'. We now consider territorial differences in poverty at different spatial scales (Philo, 1995; Hudson and Williams, 1995). Policy and ideology have long focused on urban poverty (section 2.5). But in contemporary Britain a fifth of all **rural** households live in poverty, only slightly

Figure 3.2 People queuing for free butter, Meadowell, Tyneside, 1990. The EC/EU
welfare state for farmers had produced a large butter mountain, part of which
was given away in 1990 (source: Steve Conlan).

below the national average (Shucksmith, 2000: 1). Polarisation of incomes is
even larger in the countryside than urban areas, with severe impacts on housing
for the poor (Cloke *et al.*, 2002). The rural poor suffer particularly from lack of
access to jobs, services and social interaction (Cloke and Little, 1997).

Britain has long had substantial **regional** differences in wages and
unemployment, due mainly to the fortunes of industries in world competition.
In the last 30 years regional income differences have widened further; the
South East's lead over the northern region grew between 1976 and 1993 from
20 per cent to nearly 50 per cent (Mohan, 1999: 102). Poverty is regionally
uneven: for example, in 2000 the North West and the North East had just over
17 per cent of the national population but four-tenths of those wards that
are in England's poorest 10 per cent. Nearly a third of their populations live in
those wards (SEU, 2001a: 13). These regions have mortality rates that lag
behind the richest regions by 40 years.

While the cause of differences between regions is mainly the geography
of employment, at smaller spatial scales it is a blend of employment, housing
and transport. Even the richest regions have localities and neighbourhoods
with acute poverty (Allen *et al.*, 1997), and inequalities within all regions are
growing. The **cities** have severe problems of unemployment and poor quality
jobs for the low skilled (Turok and Edge, 1999). They have higher proportions
of their population in poverty and greater social polarisation than the rest of the
country. In 1991, 71 per cent of people living in the poorest 5 per cent of wards

were in industrial-urban areas, which only account for a third of the total population (Lupton and Power, 2002). How severe city poverty can be is indicated by the case of Glasgow, where a third of the population was claiming benefits in 2003. But even in the prosperous South East, among the 3 m inhabitants of Inner London in 2001 poverty affected over half the children and over a third of pensioners (GLA, 2002). But there are also areas that have suffered calamitous decline; in the now defunct British coalfields, for instance, estimated 'real' unemployment rates in 1991 were almost double the official rate at 22.5 per cent (Beatty and Fothergill, 1998: 137).

At the smaller scale of **local authority districts** (150,000 to 250,000 population), the 88 most deprived contain 40 per cent of the national population but over 80 per cent of the poorest wards, 70 per cent of all people from ethnic minorities and an over-representation of children and disabled people. Some districts can be considered almost completely poor: London's Tower Hamlets has 80 per cent of its children on benefits. Nearly all 'failing schools' are in these poorest districts, while 40 per cent of crime is concentrated in 10 per cent of districts (Mohan, 1999: 132). The average mortality rate for the under 65s in the least healthy districts – a million people in all – is three times that in the most healthy (Shaw *et al.*, 1999).

However, even in poor districts there is considerable variation between **wards** (about 15,000 people). Indeed, at this scale there has been increasing concentration of the poor in the last two decades (Lupton and Power, 2002). In some Glasgow wards 50 per cent of men were unemployed in 1991 (Shaw *et al.*, 1999: 49). In 1998 in the 10 per cent most deprived wards 44 per cent of people rely on means-tested benefits (double the national average) and 60 per cent of children live in such households (three times the national average) (SEU, 2001a). At a still smaller scale, the poor are also increasingly concentrated in **neighbourhoods** with particularly bad housing, including council housing, with some evidence of rapid polarisation in recent years (Noble and Smith 1996). Half the children eligible for free school meals are concentrated in just one fifth of the schools (Palmer *et al.*, 2004: 52).

There are, then, parts of cities and towns which are entirely poor and where the poor are physically isolated. Nevertheless, most of the poor do not live in poor neighbourhoods or wards: for instance, two-thirds of poor children live outside the 20 per cent most disadvantaged wards (Flaherty *et al.*, 2004: 157). One cannot identify poverty simply with territory.

The *significance* of this geography is far from clear. Is spatially-concentrated poverty worse than diffuse – and worse for whom? We investigate these questions in Chapter 6 and Part III.

3.7 CONCLUSION

Low income is at the centre of deprivation, and it is largely created by (spatially uneven) labour markets and the state's mediation of them through benefits and

4

JOBS, THE ECONOMY AND SOCIAL EXCLUSION

4.1 INTRODUCTION

In this chapter we examine the political–economic relations that cause poor jobs and unemployment. Poor jobs not only give poor incomes but also disempower and devalue workers within the production process, with negative impacts on their autonomy, personal development and self-esteem. Moreover, the structures of the economy cause competition between workers for jobs, particularly at the lower end of the labour market, and thus social and political atomisation. The economy oppresses the poor not only materially but also culturally. These problems are not just the result of the competitive failure of particular territorial economies but are systematically produced by capitalism, particularly in its modern form. Section 4.2 shows the systemic ways in which capitalism generates poor jobs, economic insecurity and unemployment. Section 4.3 discusses how these are distributed unequally both socially and spatially, and how these unequal distributions have their own political effects. Section 4.4 describes the worsening of the causes of poverty, which has occurred since the economic turning point of the 1970s. In section 4.5, we offer an explanation of this period as one of economic crisis compounded by the neoliberal strategy adopted by capital, within which spatial mobility plays a central role. More generally, we shall argue that economic disadvantage is not merely patterned by space but actively produced through it.

4.2 CAPITALISM, POOR JOBS AND UNEMPLOYMENT

The disciplining of labour

For a capitalist economy to function continuously, the general level of wages cannot for long periods rise above the level which leaves business with an adequate rate of profit on capital invested – although the definition of 'adequate' is place- and time-dependent. If the rate of profit is too low, investment will fall because of lack of funds and the incentive to invest. Thus, there is a permanent downward pressure on wages and a disciplinary pressure on workers: if workers do not work sufficiently hard and/or do not accept low enough wages to make their employer's business profitable, they risk losing their jobs. This pressure acts on all workers in each industry, firm or workplace. It also acts on the unemployed, requiring them to accept jobs at a wage level that will ensure their prospective employer's profitability.

If workers do not bend to these pressures, their workplace could become uncompetitive, their firm decline, their industry collapse, the local economy weaken or the national economy of which they are part begin to slide in relation to others. This endemic competition means that other employees are competitors for jobs and promotion, and employees of other workplaces and firms are rivals. Industries where the workers have 'privileged' conditions are put under pressure to lower them to the average; and workers in each locality or nation can easily come to perceive those in other territories as potential thieves of their jobs (Gough, 1992; 2003a).

Poor jobs

These processes put constant downward pressure on employment conditions. But wages, employment benefits, security of employment and managerial control are highly differentiated between economic units – industries, types of work, territories, and firms: unequal development is endemic to capitalism. Thus **territorial industries or firms** with high levels of fixed investment often obtain higher than average profit rates. High levels of innovation in products or processes enable technical or design rents to be reaped by industries and firms. High profit rates permit workers to bargain for higher than average wages, for relatively secure employment, and for relatively good conditions of work. The immediate interest of any group of workers in a profitable area is to defend their superior economic interest vis-à-vis other groups. Conversely, territorial industries or firms with low profitability tend to compensate for this by depressing wages and conditions and imposing 'numerical flexibility' (variations in hours, times of work and number of jobs) on their workforces. High rates of unemployment weaken workers' collective and individual bargaining power and enable employers to reduce wages and intensify work. Slow rates of growth can have a similar effect.

Job skill also affects employment conditions. Workers in low skill jobs where management can directly dictate the form of tasks and the pace at which they are performed are in a weak bargaining position. They can be replaced easily as the firm is not dependent upon their particular skill or experience. A cooperative attitude and commitment from the worker is of small importance; the firm can therefore offer poor wages and conditions with impunity (Friedman, 1977). Indeed, where workers are strongly organised employers often seek to develop technologies that supplant human skills (Noble, 1984).

These processes, then, systematically produce inequalities in jobs. The general processes of the discipline of labour are intensified in particular sectors and sites. The outcome is a large 'secondary' market in jobs with low wage levels, high job intensity, and/or high job insecurity – what we shall call 'poor jobs'. These jobs not only offer low and insecure incomes, but also are often unhealthy. Typically, they blunt people's creativity, belittle them, and fail to develop their skills. These workers are denied even a limited degree of autonomy and responsibility.

Self-employment, the informal economy and criminal work

Skilled white and blue-collar workers in **self-employment** may earn decent or high wages, although their work is inherently insecure. However, low-skilled self-employment, or where the required skills are widespread, is typically not only insecure but also low paid. In the modern economy much low skill 'self-employment' is de facto employment by a firm, which thereby avoids regulatory obligations. The firm also benefits from the disposability of this form of labour, and can even make 'self-employed' workers responsible for their own costs for working. Self-employed status in these adverse conditions is a sign of particularly strong subordination to capital: the insecurity of the job causes its self-employed status rather than vice versa. Another large category of self-employment, domestic service, is also de facto employment, though in this case by an individual rather than a firm. Overall, then, poor quality self-employed jobs, while not formally capitalist, are shaped by capitalist dynamics.

High rates of unemployment or the availability of only poor jobs force people to seek their income from the **'informal economy'**. Workers, like the self-employed, may choose an informal job status to avoid tax, but run the risk of prosecution and a threat to their welfare benefits. Informal status is usually found in sectors with intense price and cost competition, where some firms survive by cutting their costs through avoidance of tax and regulatory enforcement of wages, work conditions and layoffs and the payment of job-related benefits. Thus, informal jobs typically have particularly low wages, poor working conditions, and high insecurity as both a cause *and* consequence of their informal status.

The breaking of employment, welfare and tax law in the informal sector shades into **criminal work**. This has always and everywhere been an important sector within capitalism, albeit played by partly non-capitalist rules – control of property is regulated by violence rather than by the law. Many criminal sectors are organised by pseudo-capitalist firms, whose long-term development is similar to the mainstream economy. Like mainstream firms, they are class-divided. To make a decent or high income usually requires one to be a crime boss or highly skilled; for the criminal proletariat incomes are low and unpredictable. Thus the contemporary illegal-drugs industry 'has involved "deskilling" and the growth of the "mass labour market". The mass labourers in the drugs market may be as disadvantaged in this as in any other form of employment. Mobility is limited and casual labour predominates' (Croall, 1998: 266). The great majority of burglars and robbers are young men with few conventional *or* criminal skills; compared with the smaller number of professional criminals, they face higher risks of detection and reap lower rewards (Croall, 1998: 229). The gender division of labour in crime-as-work is as sharp as in any other industry, and similarly results from wider gender roles: theft by women is overwhelmingly shoplifting, and reaps lower returns per theft than burglary. The violence involved in criminal work such as drug dealing and protection rackets renders it even more dangerous than ordinary poor jobs.

Both informal work and the majority of criminal work may thus be regarded as the extreme end of a continuum of poor jobs which begins in the formal sector – low paid, insecure, dangerous and highly market-dependent. Whilst breaking certain capitalist rules, these forms of work are reproduced systematically by capitalism: workers are forced into them through lack of decent formal jobs, and cost-cutting sectors are informalised to cut costs further. These parts of the economy, so central to the reproduction of poverty, should not, then, be regarded as outside the logic of the mainstream.

Unemployment

Unemployment is a systemic but highly variable characteristic of capitalism. Capitalism offers no guarantee or right to employment. Intensification of work and, especially, application of new technologies constantly displaces labour that may or may not be re-employed in other sectors. Population growth, change in the age structure, change in household divisions of labour, and changing needs for money income can all increase the number of those seeking employment without any corresponding increase in its availability.

Joblessness leads directly to poverty for the individual. Even in the developed countries, where state benefits are paid to at least some of the unemployed, the rate of these benefits is set at a level to encourage recipients to look for work. The least skilled are most likely to suffer unemployment, since employers, even for poor jobs, select those with higher skills and better work experience.

Unemployment is a key means by which employers create and recreate the poor jobs described above, by threatening workers with replacement. The spatially- and temporally-uneven incidence of unemployment – capitalism's territorial booms and slumps – means that in particular places and times, high levels of unemployment can enable employers not merely to make incremental adjustments in wages and work intensity but to change the rules of the game and thereby facilitate major shifts in employment practices. Unemployment, then, worsens poverty through more channels than simply the low incomes of the unemployed themselves.

4.3 THE SYSTEMATIC UNEVENNESS OF POVERTY CREATION

Social differentiation

The movement between unemployment and employment, and between poor and better jobs, is strongly constrained by **individual work history**. The long-term unemployed are a socially isolated group who increasingly know fewer and fewer employed people (Morris, 1990). Even in the developed countries with their elaborate systems of education and training outside work, a major part of workers' skills and aptitudes are acquired within employment itself, as are the contacts for alternative or new employment. Most workers intuitively know that it is easier to get another job, rather than get a job when unemployed. Individual susceptibility to unemployment and poor jobs is highly path dependent; confinement to these parts of the economy tends to be a vicious circle (see further, sections 6.4 and 6.6).

The norms of wages, work conditions, employment security and job type typically differ between workers of different **gender, ethnicity and age** within a particular territory at a particular time (section 3.5). These variations are constructed over the very long term by complex interactions between the jobs assigned to particular groups and their social lives. Social life produces different status, imputed need for a (good) job, and different access to skills demanded in waged work. Employers who rely for their profitability on providing poor jobs tend to employ socially-disadvantaged groups. The resultant poor incomes, low skills and job insecurity of these groups then reinforce their disadvantages in social life. There are thus vicious circles of exclusion and economic deprivation between employers' use of social groups and their position in the social realm (see further section 7.3).

These processes mean that poor jobs fall disproportionately on women, certain minority ethnic groups, people with disabilities, the young and the old. These groups also tend to have higher rates of unemployment, partly via their low skill levels or lack of continuity of employment, partly through inability to access jobs, and sometimes through ideologically driven choices by employers or discrimination by other workers. Some employers may prefer women, certain racialised groups or young people as cheap or malleable labour and consequently these groups can have relatively low rates of unemployment but low rates of pay whilst in employment. In either case, poverty of these groups is reproduced.

Geographical differentiation

The divide between poor and better jobs is, in part, a territorial one (section 3.6). The spatial division of labour created by investment flows and territorial competition allocates jobs very unevenly. Industries with high productivity and innovation tend to be spatially agglomerated into particular territories, whether localities, regions, nations or continents. Agglomeration enables the production and reproduction of a workforce with the skills and attitudes desired by employers, through training within and outside of production and through social life. Its spatial stability helps to foster cooperative relations between firms and their employees. It facilitates networking, collaboration and changing divisions of labour between firms for flows of goods, services, information and personnel. It helps suppliers of finance to develop a deep knowledge of the industry. And the state within the territory can develop responsiveness to the particular needs of the industry's firms in its provision of material and informational infrastructures, fiscal arrangements and regulation. These virtuous circles of agglomeration tend to be strongest in the production of complex goods and services which require strong knowledge generation and application and which use relatively skilled labour (Storper and Walker, 1989). In short, high value-added industry is often based on the *strong socialisation of production within a territory*. This high value-added can then be partly appropriated by workers in the form of good wages and conditions.

This logic, however, powerful though it is, is always in tension with the pressures of market discipline. Agglomeration tends to raise the price of labour and other inputs. Labour's bargaining power tends to be increased, and managers' control within the workplace eroded. Stability may make other firms or the state unresponsive or excessively demanding and, rather than innovating, firms may try and repeat previously successful solutions (Gough, 1996b). In the face of this ossification, investment may flow elsewhere – to where costs are lower, labour more malleable, other actors more responsive. These pressures are particularly strong in industries or stages of production which produce standardised goods or services, where little new knowledge is generated or deployed, and where tasks are relatively low skilled. These types of production tend to be located in low cost locations with plentiful supplies of non-skilled labour. Since these conditions can be found in many regions of the developed countries and in most of the Third World, this production is relatively footloose, moving from one location to another in search of yet lower wages, more pliant workers, or new state subsidies.

The outcome of these contradictory pressures – which act on *all* industry, albeit in varied ways – is highly differentiated spatial development. On a world scale, and focusing on the last 50 years or so, high value-added manufacturing, high level finance, business services and research and development work have remained strongly rooted in the developed countries. *Within* these countries, these sectors tend to be highly concentrated in particular localities and regions. These are usually places where the industry, or industries from which it evolved, have been located for many decades. However, other localities can lose their inherited socialisation of production: rural areas where agricultural and associated service employment have declined, regions where mining and manufacturing have collapsed, or towns where old-fashioned tourism has disappeared. In these areas, formerly successful socialisations have been destroyed

Loan Receipt
Liverpool John Moores University
Library and Student Support

Borrower ID:	21111120766126
Loan Date:	27/01/2010
Loan Time:	8:47 pm

Spaces of social exclusion /
31111011609136

Due Date: 17/02/2010 23:59

Please keep your receipt
in case of dispute

by internal contradictions or new external competition, and new forms of coherence have not been constructed (Harvey, 1989).

Capitalism often has its revenge on areas of previous worker prosperity. New investment in these areas is largely confined to standardised, cost competitive manufacturing and low-level office-based services such as back offices and call centres. The great majority of these jobs are poor. They have often been split off from higher-level work in the same firm or production chain precisely in order to separate them spatially and socially (Massey, 1984). Moreover, these localities are differentiated from each other merely because of cost and are therefore in sharp competition with each other, so that, once fixed investments have been depreciated, the firms and their jobs are liable to move on. In consequence, the unemployment created by the decline of old industries is often not absorbed by new investments. We thus have what we may call 'core' and 'peripheral' localities – though with the caveats that there is a continuum between these poles and that the pattern of advantage and disadvantage is never static (Sawyers and Tabb, 1984). The final element in this picture is services such as retailing, catering and leisure, which, because they are delivered locally to consumers, are ubiquitous. Most firms in these sectors design their jobs to be low skilled and 'numerically flexible', and offer low wages and poor conditions and security (Bryson *et al.*, 2004).

Core and peripheral areas create poverty in distinct ways. In the **peripheral areas** the majority of jobs are poor. Substantial levels of unemployment are chronic. Even relatively privileged sections of labour are threatened by the flightiness of peripheral types of production and, because most jobs are designed to be relatively low skilled, their replacement by others threatens most workers. It is hard for unions to be active, or even to recruit, in this environment. The informal and criminal economies usually form a large proportion of economic activity. If this situation persists over decades, as it often does, the dominant cultures of work are stamped by it. Expectations regarding wages, conditions, skill and career are low, sometimes zero; the self-confidence of individuals as economic agents is minimal. Economic exclusion thus affects the majority of the population in these areas.

The **core areas** also create poverty, however. Consumer service jobs, whether in the private or public sector or in domestic service, are as poor in these areas as in others and, driven by the high incomes of the core sectors, these tend to form a higher proportion of all jobs than in poorer regions. The markets in land and property, dominated by core production activities and core workers, raise the cost of living; those in poor jobs may lack the bargaining power to raise their wages to meet these higher costs, and may thus be even worse off in core regions than in peripheral ones (Fainstein *et al.*, 1992; Sassen, 1991; Hamnett, 2002). Furthermore, social and spatial segmentation of labour tend to be an even greater problem in core regions than peripheral ones. A relatively low aggregate unemployment rate across a region may disguise high rates of unemployment and underemployment for those with a poor work history, for ethnic minorities, and for people living in stigmatised neighbourhoods. Migrants naturally try to settle predominantly in core rather than peripheral regions. But the work of these migrant communities, at least for several decades, is typically in the consumer service sectors, in the informal economy, or in businesses owned by people of their own ethnicity; the latter may form minority-ethnic enclave industries within which bonds of ethnicity are used to create particularly poor jobs (Kakios and van der Velden, 1984; Waldinger and Lapp, 1999).

Contrasts and divisions in life chances thus tend to be particularly large in core regions. Economic exclusion appears as internal rather than external to the region. This has important ideological consequences. Whereas in weak regions economic exclusion can be the majority culture, a widely shared experience represented in dominant discourses, in core regions disadvantaged workers and the poor appear as anomalous because they have somehow failed to share in the 'general' prosperity; the cultures of the poor tend to be more suppressed, and indeed seen as threatening, within the region's dominant discourses (Stedman Jones, 1971; 1983). Schematically, we may say that, whereas in the weak regions capital mobility undermines collective organisation, in the core regions divisions *within* labour undermine collective organisation (Gough, 2003a).

The rich and the poor

So far we have considered the relation of capital to the poor as employers and as flows of investment. But it has another role: a small minority of the population reaps large incomes from appropriating a share of profits (dividends, interest, rents) and from appreciation of capital assets; they include the idle rich and corporate executives. Their incomes exacerbate poverty in the sense that they have no productive function and could therefore be reallocated to the poor without penalty. In popular ideology, the rich may be seen as undeserving and their wealth obscene, inspiring radicalism; alternatively, the media's portrayal of their seductive glamour and life styles may legitimate gross inequality. The common perception that the rich are responsible for poverty is true in these senses; but in our view the major culprit is the dynamics of investment and labour management organised by firms.

4.4 THE DEEPENING OF EMPLOYMENT-RELATED POVERTY IN THE PRESENT PERIOD

Increasing unemployment and poor jobs

In the last 30 years or so — what we shall refer to as 'the present period' — the processes of poverty creation have intensified. This has occurred in all the MDCs, though with important variations (section 7.4). It has been propelled by a crisis of profitability which became palpable in the 1970s, to which employers have reacted by holding down wages, casualising employment, intensifying work,

Figure 4.1 Disabled by neoliberal work. Colin Blenkinsop lost his foot while working on a temporary contract as a refuse collector where safety boots were not supplied (source: Richard Grassick/Amber).

laying off labour, and attacking trade unions. This has produced widening income differentials and, for some sections of the poor in some countries, absolute declines in income (sections 3.2 and 3.3).

The most obvious, and the most important, change in the present period has been a rise in **unemployment**. Whereas the 1950s and 1960s were periods of low unemployment by historical standards, since then there have been high levels in all major developed countries. *Long-term* unemployment has risen even faster, as unemployment ceased to be 'frictional', particularly for disadvantaged sections of the workforce (Nickell, 2004). The proximate causes of the rise in unemployment have been a slower rate of output growth than in the post-war boom and an increase in the proportion of the population seeking waged work, only partially compensated for by a slower growth of labour productivity. In Britain, unemployment began to rise in the 1970s, rose sharply in the early 1980s recession, and declined substantially only in the 1990s expansion. The number of jobs increased by 3.5 million between 1974 and 2004; but because of the expansion in the number of those needing a job, in 2001 12 per cent of the working-age population wanted employment. Many of the latter were not classified as unemployed, however: over those three decades there was an increase of 1.6 million in those claiming long-term sickness benefits, much of which is disguised unemployment (Webster, 2002; see further section 5.7).

Over the same period, there has been an increase in the proportion of formal employment in what we have called poor jobs. **Job security** has declined

(Doogan, 2001; Nickell *et al.*, 2002). An increasing proportion of employment is on short-term contracts. The incidence of compulsory redundancy has increased in nearly all parts of the private sector and in some state-owned industries. Insecurity of employment, which has always been high among small firms, has spread to large firms. The proximate causes of these redundancies operate at various levels. At the level of national economies there has been a slowdown in output growth and reduced rates of profit; in many mature sectors a stagnation in output combined with productivity rises from intensive fixed investment or intensification of work; at the level of firms, increased uneven-ness of profitability, with many firms suffering losses or low profits; and at the level of workplaces, increased use by management of hiring and firing to meet fluctuations in output. Those made redundant have often had to find employment in a different sector, thus destroying the value of their skills.

Another growing form of insecurity arises from employers forcing employees to become **nominally 'self-employed'**. One form of this enforced 'self-employment', with its own particular geography, is **home working**, which has expanded in manufacturing and services. The National Group on Homeworking currently estimates that there are one million homeworkers in the UK, the majority low paid rather than professional. Homeworking offloads overhead costs onto the worker, and, most importantly, isolates them from others.

There has also been an increase in **part-time work**. This has been partly powered by management's wish to casualise jobs and cut wages, given that part-time workers usually have fewer legal rights than full timers and are less likely to be unionised. It is has also arisen from decline in the absolute number of jobs in manufacturing in the developing countries and an increase in business and consumer service work: while most manufacturing jobs are full time, many service jobs are part time in order to fit with patterns of demand. Part-time work has suited some workers, especially those with dependents and some approach-ing retirement. But other part-time workers would prefer to work full time, and they suffer from lower total wages and inferior employment rights.

The imposition of **unsocial hours** has increased, partly to introduce more shifts so as to amortise fixed investments more rapidly (particularly in manu-facturing), and partly to fit service employment more closely to the timing of (socially-created) customer demand. In Britain in particular, many full-time workers, mostly male, work a large amount of overtime (exacerbating unemployment); for manual workers the aim is increased income, while for professional workers it is usually to complete assigned work, often unpaid. Hours worked are becoming more **variable** week-by-week, tailored to demand; 'zero hours' contracts are increasingly common in both manufacturing and retailing in which employers guarantee only a minimum yearly total of hours of employment and may lay off workers for days or weeks. An increasing num-ber of workers therefore work hours that conflict with social life and caring responsibilities.

In the present period, the proportion of **low wage** jobs in total employment has increased, as has the proportion of high waged jobs (Gosling *et al.*, 1994).

In Britain in the 1970s a low paid employee had a 4 per cent chance of being in poverty; by 2000/1, this had risen to 14 per cent (Millar and Gardiner, 2004). Growth in low wage jobs has been particularly spectacular in the US: average real wages for *the bottom 80 per cent* of waged employees fell by 16 per cent between 1973 and 1993 (Sklar, 1995b: 17–18). These trends are partly a result of widening differentials in wages for *given* jobs: in Britain real wages in low status jobs have stagnated or fallen, medium wages have increased, and high wages have increased rapidly. The shifting spectrum has also been due to *sectoral changes*: an increase in professional and managerial jobs; a decline in the number of jobs traditionally paying medium wages, particularly male manufacturing jobs; and an increase in jobs traditionally paid low wages, particularly in consumer services. The distribution of income (as distinct from wages) has been further skewed by large increases in the income of the rich from dividends, interest and asset appreciation, the most publicised, though not most important, aspect of which has been the enormous increases in rewards for corporate executives (Sklar, 1995b: 9).

A crucial, and often overlooked, aspect of job deterioration is the **intensification of work**. Employers have sought to increase output per worker-hour by speeding up work, cutting breaks, and requiring flexible transfer between tasks. This has occurred in both highly mechanised sectors, where the machinery may be used to pace work, and in low-mechanisation sectors where work pace is imposed through supervision and sometimes output-related pay (Gough, 2003a). Employers' imposition of flexible hours, discussed above, is also aimed at employing workers at maximum intensity. These methods were pioneered in manufacturing, but have now been applied across private and public services (Nichols, 2001). They are sometimes dressed up in a language of 'work enrichment' and worker involvement; the supermarket chain Asda welcomes its workers to 'the Walmart family'. But flexible tasks and consultation with workers are used only to the extent that they intensify work (Smith, 1994; Harrison, 1997; Moody, 1997). This regime has been applied to medium- and even high-status workers. Its imposition in low-status and low-skill jobs has been particularly brutal. Older and less-strong workers in particular often cannot stand the pace of these jobs and may have their health injured or become disabled through them (Garrahan and Stewart, 1992).

High levels of unemployment and deterioration in the security and quality of jobs have been connected, as both cause and effect, to a **weakening of the trade unions** across the whole economy with the partial exception of the public sector, a trend in all major countries (Bamber and Lansbury, 1998). This has been partly affected by sectoral shifts, particularly the decline of mining and manufacturing and rise of service employment: private sector services have long had relatively low levels of unionisation, and most of the new and fast-growing service sectors have not (yet) been substantially unionised. But it is mainly due to employers becoming more hostile towards collective action and union organisation itself. The redundancies, deterioration of conditions and wage restraint which have characterised the period have sometimes been resisted by unions, but in most cases this resistance has had meagre success, and some major strikes have been crushingly defeated, with help or direct involvement from the government. These setbacks have been demoralising for workers contemplating any type of resistance to their employer, however modest; membership of unions has seemed increasingly pointless, as their effectiveness has apparently diminished; and conservative union leaders have found increasing legitimacy for blocking effective action. This quantitative and qualitative weakening of the unions has then in turn made it easier for employers to shed labour and to worsen wages and conditions.

In most countries, **the state** has facilitated these trends. Regulations governing layoffs, training and pensions have become more permissive. Small firms have gained more exemptions from regulation. State policing of remaining employment law has become more lax. In some cases – of which Texas under Governor Bush is the most notorious – regulation of health and safety and industrial pollution has been weakened. In Britain, legislation has made strikes more difficult to organise, and solidarity action impossible. Indeed, the Blair administration has boasted that the labour market in Britain is the most flexible in Europe.

The formal economy has, then, changed in two key respects. First, there has been a relative and absolute increase in 'McJobs' – low paid, low skill, insecure, and with little or no promotion prospects. But second, this has been a part of deteriorating conditions affecting *the majority* of the workforce, albeit unevenly: increased insecurity, casualisation, weakened unionisation, unsocial hours, downward pressure on wages, and intensification of work. The employment conditions of the poor are thus a condensation of trends across the whole economy (Tilly, 1996).

Paralleling these trends in the formal economy has been an increase in **'marginal' forms of employment** including domestic service, the informal economy (that is, unregistered with the state), sex work and blue-collar criminal work. Schneider has shown that in all OECD economies the informal economy grew over the period 1989/90 to 2001/2; in the UK the increase was from 9.6 per cent of GDP to 12.5 per cent (2002: 20). Section 4.2 suggests that people have been pushed into these jobs by the rise in unemployment and by the increase in poor jobs in the formal economy. Ironically, the rocketing incomes of the rich and the increasing disposable income and declining free time of the employed upper middle class have also played their part: one in three of dual-income households employ domestic workers and there are now the same number of 'servants' in Britain, 4 million, as there were in the 1930s (Work Foundation, 2004: 4). Taken together with the increase in waged employment in consumer services, one might say that there has been a large expansion of the poor working to service the increasingly affluent middle and upper classes (Ehrenreich and Hochschild, 2003). None of these marginal jobs has any security. In many the workers are subjected to violence, sexual harassment and threats to health. Nearly all produce very low incomes. Most of them are socially and physically isolating – the employer's house for the domestic worker, the home worker's own house, the lone entrepreneurship of the person attempting to make a living out of the informal economy, sex work or property crime, the individualising of the fictitiously self-employed (Williams and Windebank, 1998). But the increase in these informal forms of work is congruent with the increase in poor jobs in the formal economy.

Increasing social unevenness of employment

In the present period the **gendering** of working-class employment has changed sharply. Male employment rates have tended to decline because of cuts in jobs in many branches of manufacturing, primary industries and utilities. At the same time, sectors employing high proportions of women such as consumer services, public services and routine office work have expanded. Thus, while male rates of employment have fallen in most countries, especially for the lower skilled, those for women have tended to increase. But these new opportunities have been largely in poor jobs and largely part time, though there have been some inroads by women into better jobs previously reserved for men; and the increasing number of poor jobs have been filled disproportionately by women, an increasing feminisation of the bottom of the labour market. Stone (1997) has calculated for the North East of England that whilst the number of jobs for women has increased in the present period, the total wage income paid to women workers in aggregate did not increase at all.

These deteriorating employment conditions for working-class men have required their female partners to seek (more) waged work. Moreover, as lower wages have stagnated, two wages have become more necessary for couples and families. Couples with only one wage income are more likely to be poor and the poverty of those with no wage income is accentuated. Both two-wage *and* no-wage households have become more common in the present period.

These processes have also altered participation in waged employment by **age**. People over 50 are increasingly discriminated against on the grounds that they will not, and sometimes cannot, increase their intensity of work as required by employers, and are leaving employment to avoid intolerable conditions. Having been absent from waged work for periods of childcare additionally disadvantages many older women. Young people are in some cases preferentially employed on the basis that they can be paid less and offer less resistance to management diktats due to lack of work experience. But in other cases, particularly in countries such as France where youth cannot legally be paid lower wages, employers favour older workers and youth unemployment is high.

The employment position of most **BME groups** has worsened in the present period as employer-organised competition with white workers has increased. The most vulnerable group of all, **illegal immigrants**, appears to be increasingly drawn on by sweated sectors. Many are employed through gang masters, who take a large cut of the wage paid by the employer, charge exorbitant rent for accommodation, organise dangerous work, and react to complaints with violence or threat of disclosure (Lawrence, 2005). Some are forced to work for nothing in order to pay off debt to people-smugglers. Some, such as those forced or tricked into sex work, are effectively enslaved and subjected to chronic violence. The extent of such barbarity reveals the weakness of state surveillance of employers' work practices, a result of deregulation in the name of the 'flexible labour market'.

Increasing spatial and temporal unevenness of employment

Geographical differences in employment have widened in the last 30 years. Differences between MDCs have become more pronounced, particularly in their impact on the poor. In the 1970s and 1980s the countries which had been competitively weakest during the post-war boom were the worst hit by a slow-down in output growth and rises in unemployment; these included Britain, southern Europe and, in certain ways, the USA. During the 1990s, however, these economies strengthened their competitive position, while the countries that had had the strongest growth during the boom, (West) Germany, France and Japan, experienced deepening stagnation and rising unemployment.

Differences also increased between regions within countries (Green and Owen, 1998; Hudson, 1988; Hudson and Williams, 1995). Many old industrial regions, and even more sharply, particular towns and villages, suffered major contractions in manufacturing and mining employment; employment in agriculture and fishing declined rapidly in many areas. Employment growth tended to be concentrated in city-regions specialising in the growth sectors of the present period – finance, high-level producer services and the media; in localities specialised in complex manufactures; and – with poorer quality jobs – in localities with mass or niche tourism. Divergences between localities in both quantity and quality of jobs increased sharply.

The deterioration of employment has also been highly **uneven in time**. Capitalism is marked by cycles of growth and contraction, including 'long waves' of roughly 50 years' duration and 'business cycles' of five to ten years (Mandel, 1978). In the phase of stagnation of a long wave such as the world economy has experienced since the 1970s, extensive investment is weaker and unemployment is higher than in the phase of expansion. The present period has been marked by business cycles of far greater amplitude than was the case in the post-war boom. There were sharp downturns in 1973–6, 1979–82, 1989–92 which hit all major countries at more or less the same time; since the early 1990s the Japanese economy has been in stagnation, and since 2000 continental Europe has been hit by recession. These downturns have seen the scrapping of capacity, particularly in manufacturing but latterly also in business services, and sharp rises in unemployment. Unemployment in Britain, for example, rose from 4 per cent to 12 per cent in the 1979–84 recession. Moreover, these downturns have both forced and enabled employers to intensify work, change conditions of employment and hold down or cut real wages and pension entitlements, using the threat of redundancy.

Thus the deterioration in the job market as a whole has fallen disproportionately on already-vulnerable social groups and particular geographical areas, and has been most rapid and coercive during cyclical downturns.

4.5 A LONG CRISIS OF CAPITALISM

The profit crisis which sparked these attacks by employers on working conditions was the culmination of a steady reduction through the period of the long boom in the world average rate of profit; by the early 1970s this key regulator of investment and industrial relations had fallen to historically very low levels. Low profits have resulted in low rates of extensive investment and slow long-term growth over the last 35 years. The reasons for this wave of economic stagnation are complex and much debated. In our view, they do not stem from exhaustion of a Fordist production paradigm or regime (cf. Clarke, 1992; Gough, 1996a), but rather from two fundamental and necessary aspects of the long boom.

First, the capital invested per worker increased over the period of the boom, not only because of a tendency to increased capital intensity in each sector but also because of greatly increased investment in research and development, in transport, in state subsidies to industry, and in public services reproducing the labour force. This increase in the capital invested per worker put downward pressure on the rate of profit (Mandel, 1978; Armstrong *et al.*, 1991: chapter 14; Duménil and Levy, 1993). Second, the boom was sustained by relatively consensual relations between capital and workers and between sectors of capital. But, as the boom proceeded, these came under strain and erupted in conflict. Relatively low rates of unemployment encouraged workers to make stronger demands on their employers and the state, and union density increased. People were emboldened to demand better public services and urban environments. Women, whose social position had changed rapidly through the boom, demanded real equality and the resources to support it. Black people, most decisively in the USA, demanded an end to the disparity between rising average living standards and their own marginalisation in employment, housing and services. At the same time, firms sought ever-greater forms of subsidy and support from the state.

These varied demands tended to put further downward pressure on profit rates, directly or through taxation. However, the falling rate of profit in turn made both firms and the state more resistant to demands that would increase their costs. The result was a generalised upsurge of social conflict from the late 1960s through the 1970s in all the developed countries, with the notable exception of Japan where the average profit rate remained high. There was a wave of militant trade union actions in defence of wages and jobs. The civil rights movement in the USA evolved into a politically heterogeneous but forceful black movement, which won considerable resources from the federal government to address urban poverty. The women's, lesbian and gay movements were reborn. Struggles broke out around a range of urban issues – housing, public transport, and the environment (Dear and Scott, 1981; Castells, 1982). These varied organisations and struggles, though separated by social composition, traditions and political focuses, encouraged each other. The outcome was a wholesale *politicisation* of economy and society (Habermas, 1976; Offe, 1984). The naturalness and fairness of markets and the priorities of business were called into question.

Capital reacted to this crisis by (increasingly) adopting a neoliberal strategy. Society was to be depoliticised by imposing market disciplines on workers and citizens and on firms themselves. Profit per worker was to increase sharply. State spending and thus taxation was to be reduced. The welfare state and statutory protection of workers' conditions were to be cut back. The state pursued 'sound money' and deflation (though inconsistently), and this played a key role in depoliticising the economy (Clarke, 1991; Bonefeld *et al.*, 1995). Demands from unions, residents' organisations and the social movements were delegitimated by the reality and ideology of increased competition, the imperative of increased productivity, and the 'need to control inflation' (Deakin and Wilkinson, 1991). The state's actions in deregulating employment and weakening unions were central in propagating the idea that workers can achieve economic security only through their individual effort and mutual competition.

In response to the low average rate of profit, the overall rate of investment fell and investment became more focused on raising productivity and shedding jobs rather than expanding capacity. This resulted directly in increased unemployment, with a further negative feedback on employment via weakening demand. As described in the last section, employers sought to increase profit per worker through wage restraint, work intensification and numerical flexibility. These measures faced opposition from the better-organised trade unions, especially in mining, manufacturing and the public sector. Management and the state accordingly became far more aggressive and intransigent towards unions and industrial actions than had been the case in the long boom. This offensive has been successful in numerically weakening unions, undermining workers' confidence to resist and their sense of collectivity.

This offensive of business was easier to implement at the lower end of the job market where the threat of unemployment was greater. Indeed, management was often tactically compelled to give wage increases in occupations with acute shortages of skilled labour. Moreover, firms in labour intensive sectors lacking technical rents, such as most consumer services, faced increasingly sharp cost competition while for technical reasons finding it hard to increase productivity via mechanisation; they were thus under especially great pressure to attack wages and working conditions (Gough, 2003a).

Neoliberalism has exploited, and sought to enhance, **international economic flows** as a means of imposing market discipline on workers and firms. Capital movements between developed countries have increased: corporate takeovers and mergers; investment to win access to skilled labour or final markets; international loans to business, consumers and governments; and speculative capital flows into financial assets and property. Certain types of production have been substantially shifted from the developed countries to newly industrialising countries to exploit cheaper labour: the manufacture of standardised and bulk products, low skill labour-intensive manufacturing, and repetitive, narrow-task office work. This increased capital mobility is not an inexorable result of 'globalisation' but is a strategy by individual firms to increase their profitability, and a strategy by business as a whole aimed at increasing its power over labour (Albo, 1993; Broad, 1995).

The states of developed countries have weakened controls on the flow of money capital across their borders. Orchestrated by international regulatory bodies such as the World Trade Organization (WTO), states have lowered tariff and non-tariff barriers to trade (though expanding export subsidies), thus exposing enterprises and workers to geographically wider competition (O'Neill, 1997). Cross-border productive investment and ownership restrictions have been reduced, in particular by opening up previously state owned services to private tender, a process now also driven by the WTO. While capital has thus been enabled to be more mobile, controls on the movement of labour across national borders have largely been maintained. However, substantial immigration from the Third World to some developed countries has continued in order to swell the labour force for poor jobs and to meet shortages in some skilled jobs. Immigration controls have been maintained partly to placate public opinion (sections 9.2 and 9.4); they have the effect of creating a super-exploited work-force of illegal migrants (section 4.4).

MDC states have pursued these forms of international deregulation in congruence with their internal deregulation, although with considerable pressure from the most powerful state, the USA (Gowan, 1999). Neoliberal reforms made by Third World states, however, have in many cases been done under economic, political and ultimately military pressure from the MDCs, both bilaterally (particularly by the USA) and via the WTO, World Bank and International Monetary Fund.

While the neoliberal actions of business and states have had much success in depoliticising the economy, they have not so far succeeded in restoring the dynamic accumulation of the boom years. Average profit rates at the end of the 1990s boom were no higher than those in the late 1960s, and an increasing proportion of profit has been appropriated by financial capital (Duménil and Levy, 2001b). Historically-low profit rates have resulted in sharp economic cycles, as expansions have been pumped up by stock market speculation, enormous debt creation and international lending, leading to severe productive overcapacity and recessions. The resultant switchback economy impacts particularly severely on the poor (section 4.4). There is no sign of an international economic revival which might begin to improve their situation.

5.2 STATE SPENDING, TAXATION AND POVERTY

Out of all social groups, it is the poor who *potentially* have the greatest gain to make from welfare spending; and it is often thought that they are indeed the main beneficiaries. But the class politics of state spending and taxation minimise, and often negate, these benefits. Particularly since the Second World War, a substantial part of the tax burden of the welfare state has fallen on the employed poor through income taxes, and on all the poor through the taxation of consumption. The extent of this burden depends on the division of taxation between the population and business, and its distribution among the population; both of these are the subjects of continuous political conflict. The distribution of taxation *within* the population depends on several mechanisms. Income tax is usually progressive (a higher rate on higher incomes), but with much variation as to thresholds and differentials. Income tax allowances may be given for 'essentials' such as social insurance, pensions, savings or house purchase; these tend to be regressive both because they are higher for higher-rate taxpayers and because the better-off spend more on these items, and they are thus a classic component of 'the middle class welfare state'. Taxes on consumption are neutral except to the extent that there are lower rates for basic items of consumption and higher rates on tobacco and alcohol. Property taxes are usually regressive since the poor spend a higher proportion of their income on housing than the rich.

The *local* funding of welfare services provided by the local state tends to deepen its regressive nature. Local taxation to fund local services tends to thwart redistribution from the better off to the poor. Poor localities are forced to raise taxes at a higher rate to reap a given per capita income because of their low tax base. Thus, key aspects of the politics of distribution are the distribution of taxation between central and local government (and regional government, where it exists), and the right of communities to construct their own local authorities. These arrangements vary enormously between different countries. The USA has particularly regressive arrangements: in most states, a large proportion of funding for local services is locally funded, and high income neighbourhoods can easily form separate local authorities in order to avoid paying for services for their poorer neighbours. Tax rates and quality of services thus differ enormously between poor and rich localities (Gottdiener, 1987). In Britain, local authorities are both tightly controlled and strongly funded by national government. Nevertheless, tax rates tend to be higher, and public services less well funded, in poorer localities. In this respect, local autonomy harms the poor.

Overall, then, taxation of individuals is usually only weakly progressive or is even regressive. Taking into account that public services benefit the whole population while state transfer incomes go principally to the poor, the balance of taxation and benefits associated with the welfare state may not be progressive across the population as a whole, and is a smaller subsidy to the non-working poor than at first appears.

This distribution has worsened for the poor during **the present period of economic stagnation** (section 4.5). In line with neoliberal precepts, the central strategy of governments has been to attempt to **reduce state spending** as a proportion of GDP. This has been a response to pressures from business to divert income to profit and thus counter the low rate of corporate profitability (Mandel, 1978; Clarke, 1988). This has been effected directly by cuts in corporate taxation, but also through cuts in taxation of the population, which raise profitability by putting downward pressure on wages. This strategy has been contradictory to the extent that cuts in certain types of spending can require higher spending elsewhere, sometimes with a negative sum; this is true in particular of cuts in expenditures sustaining employment which result in greater spending on the unemployed and other negative social impacts (for the case of coal mining, see Glyn, 1988). But cuts in spending have not been evenly distributed between different purposes: subsidies to business, such as for research and development and exports, and spending on the military and police, have tended to be protected, while cuts have been concentrated on job subsidies, public services and state transfer incomes – spending of particular benefit to the poor.

At the same time, the burden of taxation has been shifted. In nearly all developed countries **taxation has been shifted from capital to labour**. Decreases in the tax rate on profits and increases in allowances have been justified by arguing that firms would relocate production abroad if taxation was above the international norm – a norm which, *ipso facto*, is bound to decline. Despite this race to the bottom, many firms moved their nominal ownership to tax havens, and it is estimated that a quarter of world corporate assets are now located in them. Moreover, firms have increasingly used international transfer pricing and fraud to further reduce their tax bills. But the argument of governments that 'globalisation' requires ever-lower corporate taxation neglects the rooting of the majority of world production in the developed countries, the dependence of much production on specific types of labour power, and the immobility of much service work (sections 4.3 and 4.4). Moreover, the major governments have taken no serious steps to prevent the growth of tax havens or of tax evasion. One informed judgement is that the British treasury was losing £20bn p.a. in the late 1990s through illegal evasion of corporate taxation; surveillance was reduced despite the fact that returns to spending on it were around tenfold. Reductions in corporate taxation have therefore been a *political choice*, *shared* by 'competing' states, to shift income from the state to capital. Individuals have thus increasingly funded welfare, with any redistribution taking place within the working class rather than from business to the working class.

The taxation of individuals has also become more regressive. In many countries taxes have been shifted from income to consumption, notable cases being the large increase in VAT rates in the EU and the introduction of this tax in Australia. The distribution of income tax has been shifted by reducing the highest rate of tax and by failing to increase tax thresholds in line with incomes. Rules for taxation of unearned income and the use of tax shelters have enabled the rich to minimise their taxes. The most inegalitarian reforms have occurred in the USA: Reagan's reforms in the 1980s are estimated to have engineered a transfer from the poor to the rich three times larger than the progressive redistribution of Johnson's reforms in the 1960s (Davis, 1986); virtually all of Bush's tax cuts in 2002–3 went to the richest 1 per cent of the population. In Britain, local taxation has been made more regressive: a mildly regressive property tax, the rates, was replaced in the late 1980s with the outrageously regressive poll tax (a uniform charge on all adults) which, in the face of resistance, was replaced by the current council tax, a property tax more regressive than the rates. In all countries, rates of local taxation have tended to increase faster, and services deteriorate more rapidly, in poorer areas, as local authorities have faced weakening tax bases and eroded transfers from higher levels of government; increasing spatial inequality in state funding and spending has therefore contributed to deepened social inequality. The outcome of these processes in Britain is that the poorest 20 per cent are now taxed at a higher rate than the richest 20 per cent (*Economic Trends*, June 2004).

Overall, then, the balance of welfare for and taxes of the poor has worsened in the last three decades. We now turn to the benefits of welfare.

5.3 HOW PUBLIC SERVICES FAIL THE POOR

While the poor are reliant on public services, they fail to compensate for social disadvantage, and sometimes actually reinforce it. We are concerned here with education, health, social work and social services; we look at public housing, land use planning and public transport in Chapter 6, but we sometimes refer to them here to highlight commonalities. We first examine the 'heyday' of welfare services during the post-war boom, and then subsequent changes.

The first, and most general, reason why public services fail the socially excluded is **parsimonious funding**. Even in times of prosperity there is resistance to increases in taxation from parts of business and the population. Moreover, extending state intervention into reproduction transgresses the capitalist belief that the autonomy of the individual is sacrosanct. Inadequate funding is all the more serious given that public services are labour intensive and increases in productivity are hard to make without sacrificing quality, so that their cost tends to increase as a proportion of GDP even without improvements.

A second problem of public services is that they are distorted by, and even reproduce, power relations in society. We saw in Chapters 1 and 2 that historically public services often functioned to reinforce the social order of class, gender and 'race'. The priorities and methods of delivery of the modern welfare state have, first, been marked by the **priorities of business**: to have access to a disciplined, adequately trained and healthy workforce. Thus, services for the elderly such as geriatric medicine and care homes have been poorly funded and their quality has seldom been a political issue; the same is true of services for the disabled. Rehabilitation and other services for users of hard drugs, who are usually on a short road to poverty, are derisory because they are not seen as useful labour power. Secondary and even primary education has been structured to feed young people into the labour market. This was quite open in the post-war system of grammar and secondary modern schools, the former geared to producing future professional workers (from overwhelmingly middle-class children), the latter producing manual workers (and educating the great majority of working-class children). But since the advent of the comprehensive school in the 1960s, there have been similar outcomes through the expansion of private schools, continued selection in the state sector, and streaming in the comprehensives. Moreover, a large part of the real curriculum has been to teach children to be quiet, sit still, postpone gratification, learn what they are told, and accept the teacher's authority, reinforced by compulsory religious education; these are precisely the core qualities which employers require of their employees, and they are *all* the qualities required in poor, routine jobs (Bowles and Gintis, 1976; Illich, 1983). Education in Britain has, to be sure, also included elements for the development of imagination, creativity, critical thought and self-realisation; these were strong in the 1960s when full employment removed the pressure on schools from parents and pupils to produce factory- or office-fodder. But even in the boom they were in tension with pressures to mould children as labour power.

Class in a different sense, that of **differences in income and status**, also influences public services. Public services may be better funded in middle class areas not only because of geographical-fiscal arrangements (section 5.2) but also because the middle-class has greater resources, confidence and connections to lobby local government (Whitehead, 1992: 277–8). The state also often favours the 'respectable' working class over the poor: in the post-war boom, new public housing in Britain was allocated predominantly to the families of white skilled workers (Chapter 6). Moreover, the poor tend to get less out of *given* facilities. Lacking daily experience of performing with middle-class people, they tend to negotiate less well with the teachers, doctors, housing managers and social workers. The latter for their part often see poor people as 'rough' and 'difficult', and so less deserving. The poor often *need more* from the service – more intensive teaching due to lack of middle-class cultural capital, more health care due to poorer health, more help from social workers due to material deprivation and consequent family tensions (Tudor Hart, 1971; Carrier and Kendall, 1998). Poor children often have a low attention span when they are at school due to hunger or to a highly processed diet, and have no quiet space at home to do homework (Millar, 2004).

Moreover, poor people may reject the authoritarian aspects of public services, and thus fail to 'benefit' from them. This is particularly true in that most contradictory of services, education. Poor children tend to reject the class-disciplinary aspect of education. They often do not see the kind of knowledge imparted by schools as relevant to their situation, since they accurately

taking responsibility for their own welfare and managing its risks. Redistribution of income is to be eschewed. People must rely more on their private income and savings, and, for those of working age, protection from poverty is to come from employment: 'labour' is now most definitely 'a commodity' (Carmel and Papodopoulos, 2003; Yeates, 2003; Dixon and Hyde, 2001). Three elements deserve discussion: cuts in the level and availability of benefits; pressure to use private assets, particularly for retirement; and pressure and incentives to take up low-paid jobs.

5.7 CUTS IN BENEFITS AND IMPOSITION OF FAMILY DEPENDENCE

Economic stagnation since the 1970s has produced a double squeeze on benefits: pressure to reduce state spending has come up against increased need for state transfer incomes (section 5.6). Accordingly, the Thatcher governments in the 1980s cut both levels of benefit and eligibility for them. The indexation of pensions and benefits to average earnings was abandoned; pensions were increased merely with inflation; other benefits were actually reduced in real terms, and even more relative to average earnings, over the decade (section 3.3). Eligibility for unemployment and invalidity benefits was restricted (on similar change in the USA see Bryson, 2003: 87). As a result, in the late 1980s the state was able to reduce spending on the unemployed (Dean and Taylor-Gooby, 1992: 61–2, 69–70). Moreover, insurance contributions were increased and taxation of benefits introduced. Funding for durable items for those on benefits was severely restricted; grants were largely converted to loans; and loans were increasingly refused on grounds of claimants' inability to repay them (Becker, 2003)!

These cuts forced many people into greater reliance on their family: the space of support shrinks from society to the household. Indeed, other reforms made these intensified family dependencies explicit. The Conservatives argued that long-term changes in household forms were principally due to benefits undermining family ties, enabling young people to move out of their parents' home and encouraging young unmarried women to have children (section 1.5; Murray, 1990). Accordingly, income support was removed entirely from 16 to 18 year olds and reduced for those under 25. Ironically, research subsequently showed that half of the young people denied benefits were turned out of the family home by their parents as they were too costly, often then becoming destitute and homeless. Measures were introduced to 'dissuade' women from having children on their own (again, following the USA: Rowlingson, 2003: 23). There has been increased pressure to use familial ties for financial support, for example through stepped-up policing of child maintenance payments, making it more difficult to escape unsatisfactory or violent relationships. Together with the increase in unemployment, these changes in benefits were the main reason for an increase from 12 per cent to 20 per cent in the proportion of the total population in poverty during the 1980s (section 3.2; Dean and Taylor-Gooby, 1992: 9).

To enforce these changes surveillance was intensified. The quality of information provided to claimants has deteriorated (McKay, 2003). The efficiency of administration of benefits has declined with intensification of work and wage restraint for staff and because of privatisation. Ability to attend to individual needs has been further weakened (Walker, 2001).

5.8 RETIREMENT INCOMES AND THE CASINO ECONOMY

Pushing people into greater reliance on their families has been closely linked to greater reliance on personal income and savings, what Hewitt (2002) has termed 'assets based welfare'. Cuts in public services have increased use of personal insurance (for instance for health care) or use of individuals' savings (care of the elderly). The most important example has been retirement pensions. In capitalism, retirement incomes are based on labour market position during working life, and on family – the division of work and income by couples and income from adult children; they are consequently strongly differentiated by class and gender. Class inequalities are amplified by the dependence of savings on *disposable* income, and by regressive tax rebates on savings. In Britain in the post-war period, pensions were centred on the contributory state pension and on occupational pension schemes, with a fallback of means-tested income support. Women were far more dependent on means-tested benefits, and increasingly so over time: it bears repetition that by the late 1990s 51 per cent of women but only 14 per cent of men were so dependent (Ginn and Arber, 1996). Occupational pensions were held mainly by workers in stable, full-time jobs in large firms or state employment, predominantly men. Their coverage peaked in 1967 at 53 per cent of employees, and then declined as enterprise fragmentation and insecure and part-time work increased (section 4.4; Sunley, 2000). In response, in the 1970s state pensions were improved by the introduction of widow's benefit and the earnings-related pension (SERPS).

Reforms since the 1980s have made the pension system yet more unequal (Blackburn, 1999). While reducing the relative value of state pensions, governments have encouraged 'personal pensions' funded by private savings by introducing new taxation rules and a right to opt out of occupational pensions. These are presented as suitable for the 'flexible' labour market which renders occupational pensions too 'rigid', as giving greater choice in forms of saving, and as avoiding dependency on the state – post-modern pensions. Personal pensions are also intended to increase what is argued to be an inadequate total rate of savings for retirement, especially by poorer people (DSS, 1998; Rowlingson, 2003). The Labour government set a target of changing the incomes of the retired from 60:40 public:private to 40:60. These reforms have shifted funding of pensions from current national income to investment funds (on similar reforms in other EU countries see Miles and Timmerman, 1999; Taylor-Gooby, 1999).

In recent years this strategy has been legitimated by a supposed 'demographic time bomb', an increase in the ratio of retired people to the working population over the next 50 years or so. However, expert assessments indicate that only very modest real-terms increases in pension contributions are needed on demographic grounds, which could be easily managed through public provision (Mullan, 2002). The real motive for increasing private provision is to shift the cost of retirement from the state and employers onto wage earners, thus also benefitting finance capital. The change involves a profound ideological shift: the elderly are to be seen not as a group deserving the support of society but rather as former earners who have made themselves *independent* from society through hard work and canny investment choices (Langley, 2004). In line with neoliberal precepts, provision becomes both more individual and more capital-based.

These changes have further increased inequalities. Between 1981 and 1997 the state pension fell from 32 per cent to 16 per cent of average male earnings; SERPS was cut substantially and the widow's pension halved. The current government's pension improvements have done little: the Pensions Credit guarantees only 20 per cent of average earning, and the new second state pension for many people promises no more than income support (Sunley, 2000). The government has sought to minimise competition between state pensions and personal pension plans (though the National Association of Pension Funds is still opposed to the government's measures). The take-up of personal pensions has increased rapidly, but they are even more unequally distributed than occupational ones. By the late 1990s coverage of non-state pension provision was 90 per cent for those with income over £20,000 but only 40 per cent for those under £7,000, and men had roughly twice the coverage of women (Sunley, 2000). Unsurprisingly, *contributions* vary strongly with wages, and have been lowered for many by the rapid increase in unemployment and (involuntary) early retirement of over-50s (McKay, 2003). As regards *current* pension incomes, in 1999 the poorest four quintiles of the retired *in aggregate* drew only 80 per cent of the occupational pension income of the highest quintile, and for personal pensions this figure was only 40 per cent (McKay, 2003: 196). Because of the long-term increases in payments into private schemes, *average* retirement incomes have roughly kept up with average earnings. But inequality has increased: the richest quintile increased its income by 40 per cent relative to the lowest quintile between 1981 and 1998, a difference projected to increase further in the next 20 years (McKay, 2003). Moreover, the growth of private pension coverage has reduced the political pressure to improve state pensions.

Because of their basis in investment, non-state pensions are risky and unreliable. It is very difficult for a saver to assess risks, especially people on lower incomes who lack (what can appositely be called) this 'cultural capital' (Aldridge, 1998; Leyshon *et al.*, 1998). Thus, in the early 1990s over a million workers, particularly poorer ones (McKay, 2003), were fraudulently persuaded to switch out of their occupational pensions, resulting in an aggregate loss of around £11 bn. Moreover, private pensions are *inherently* unpredictable because of their basis in capital assets (Miles and Timmerman, 1999; Minns, 2001). The collapse of world stock markets since their speculatively inflated peak in 2000 has produced a crisis in private pensions. Most firms have changed their pension schemes so that the pension is no longer guaranteed as a multiple of the final wage, and the level of the pension is therefore dependent on the financial markets at the time of retirement; the latter is true, a fortiori, for personal pensions. Occupational pensions have also been weakened by companies reducing their contributions to them, or by (legally or illegally) raiding them for funds. The pensions in which governments are urging people to invest more are thus high risk because of the vicissitudes of capital assets in pension funds, the wider financial markets, and the worker's own industry. The low rate of take-up of personal pensions by low-income people is thus rational. Governments have tried to overcome this through Individual Savings Accounts and by packaging private pensions as 'stakeholder pensions', but these too have failed to increase uptake by the poor. Working-class pensioners who own their homes are fortunate in having an asset which can be re-mortgaged for income; but they are then dependent on asset values in another switch-back market, particularly as house prices can decline precipitately in poor neighbourhoods.

The new regime of pensions is thus imposing increasing burdens on low wage earners while promising them a poor retirement. Longstanding inequalities in pensions arising from the labour market and gendered places in the family are being deepened. Retirement incomes are now dependent on 'fictitious capital' in a period when this is particularly unstable.

5.9 'WELFARE TO WORK'

The enormous increase in unemployment in Britain in the 1970s and 1980s created considerable political embarrassment. While governments and business abandoned the post-war promise of full employment, working people did not lose this expectation so quickly. In the 1980s the Conservatives went to great lengths to disguise the level of unemployment, changing its accounting in 28 respects; by 1991 this had the effect of artificially reducing the unemployment rate from 12 per cent to 8 per cent (Dean and Taylor-Gooby, 1992: 61). The unemployed were encouraged to move onto invalidity benefit, as the government was in this case willing to *increase* benefit levels for political aims; similarly, compulsory training schemes removed people from the unemployment register at small cost.

The main strategy, however, has been to force the unemployed into work, or rather, to take any work available. 'Workfare' or 'welfare to work' – borrowing terminology as well as policy from the USA – was introduced by the Thatcher government in the 1980s and has since been reinforced. While liberal Britain and the USA have taken this approach furthest (Handler and Hasenfeld, 1991), it has been influential in other countries (OECD, 1999; Peck, 2001). Waged work is presented as always superior to drawing benefits, not simply because income is higher but also because the individual achieves independence from the state. The degree of pressure on claimants to work varies with their age, health, abilities and household situation; but the principle is seen as applying to all, and has been extended ever-wider, in recent years notably to lone parents and people on invalidity benefit (Carmel and Papodopoulos, 2003). Despite this ethic of work, governments do not ensure that jobs are available for the unemployed: the state will ensure *employability* rather than *employment*.

The system for producing 'employability' centres on an insistence that claimants, after a period of compulsory counselling and help with job search, participate in normal or subsidised employment, training or community work on pain of losing benefits. Under the Blair governments, the system has been presented as being a *contract* between the claimant and the state, within which claimants have both 'rights' to training and help with job search and 'responsibilities' to improve themselves and take whatever work is offered (Department of Work and Pensions, 2002; Carmel and Papodopoulos, 2003: 41).

Workfare can be carried out in two modes: **coercive**, in which the unemployed person is pushed as cheaply and rapidly as possible into any job, and **developmental**, where they receive training in substantial skills and back-up such as child care or treatment for addictions (Theodore and Peck, 2000; cf. sections 1.3 and 1.4). The former 'work first' mode seeks to impose habits of wage work through jobs, which are themselves usually highly disciplinary, and serves businesses offering poor jobs. The developmental mode can facilitate access to better quality employment, and potentially ameliorate skills shortages. But it is more expensive to deliver and, since many of the unemployed start as 'uncompetitive' for good jobs, such programmes often appear inefficient in placing clients in jobs. Theodore and Peck argue that welfare to work in the USA has become strongly disciplinary, and in Britain only slightly less so, though there are important variations between localities. In Britain, the poor quality of training for the unemployed continues the system initiated by Thatcher's abolition of the corporatist employer–union–state training system (Peck, 1993).

The ability of workfare programmes to get people into a job depends crucially on labour market conditions (Finn, 1987). In Britain in the late 1990s, with a relatively strong demand for labour, the New Deals for youth and the long-term unemployed had some success. But even then, employers were prejudiced against claimants and had to be lured into taking them by schemes sending only 'suitable' applicants, by employment subsidies, and by the right to sack after a trial period (Bryson, 2003). There was strong local variation (Turok and Webster, 1998; Peck and Theodore, 2000): the proportion of clients going into unsubsidised jobs varied from 30 to 50 per cent between local programme areas, and even subsidised jobs were no more available in weak than strong areas (Martin *et al.*, 2003); the type of training offered also varies (Theodore and Peck, 2000). Moreover, the general quality of jobs obtained by clients of these programmes in both the USA and Britain is poor. Of *all* unemployed people going into employment in Britain, White and Forth (1998) found that 75 per cent obtained contingent jobs, that is, jobs that are designed to be short-term or temporary. For those coming from the New Deal for Young People, only 30–60 per cent were still in their job after 26 weeks, showing the high level of instability and casualisation, worst in high unemployment areas. The justification for pushing people into *any* job, that it will be a stepping-stone to better employment, is illusory (Peck and Theodore, 2000).

Coercive programmes reinforce differences in 'employability' among their clients. Successively 'less employable' clients are pushed down the hierarchy of options from subsidised jobs to training to community work. Clients may be deemed less employable because of their caring responsibilities, which are not counted as 'work' (Levitas, 1998), or because of their health problems or disabilities. Programmes tend to take the easy route to minimise spending and to meet 'performance' targets (Theodore and Peck, 2000). The *state's* schemes are thus dominated by differences created by *markets* – localities' supply of jobs and the socially created employability of clients.

The compulsion in workfare implies that many of **the unemployed are work shy or are doing undeclared work** (Mead, 1997a; Green, 1998). Public opinion on this has fluctuated. In the 1980s soaring unemployment made it hard to blame it on fraudulent claimants. But during the 1990s, as official measures of unemployment declined, the majority view again became opposed to increases in benefit levels (despite *growth* in support for increased spending on welfare *services*), and by 2000 40 per cent believed that most claimants were fraudulent (Sainsbury, 2003: 292). These attitudes legitimated the extension of coercive workfare; but they are contradicted by varied research. Fraud is not common (Sainsbury, 2003). Lower benefit levels do not decrease rates of unemployment (McLaughlin *et al.*, 1989). Dean and Taylor-Gooby (1992) reported that unemployed people are nearly all keen to work, motivated by self-esteem as much as by money; they subscribe to mainstream values and have no 'culture of dependency' (cf. Gallie and Vogler, 1990; Jackson, 1994). Conversely, disciplinary workfare programmes undermine people's motivation and even ability to take up employment. Poor training programmes do not help the move into employment (Ainley 1988). Dean and Taylor-Gooby found that 40 per cent of their respondents wished to undertake training to equip them for a skilled job but were refused it. Indeed, poor training and the pressure to take any job undermine self-confidence and the wish to use and improve *existing* skills (Dean, 1995). Thus, in the USA those whose benefit has punitively been withdrawn frequently stop looking for work (Bryson, 2003). Dean and Taylor-Gooby identified the real barriers to taking up waged employment – caring work, lack of skill, age, health, experience, the stigma of being a claimant, and the poverty trap.

Workfare programmes do not merely respond to, but also **shape labour markets**. The Department of Work and Pensions (2002) has boasted that the 'planning, provision and delivery [of programmes for the unemployed] are geared towards employers' recruitment needs'. Peck and Theodore (2000; Peck, 2001) have argued that workfare worsens the quality of jobs at the low end of the labour market. It provides employers with an already-stratified supply of labour. By lowering self-esteem workfare makes people more willing to accept poor conditions. It encourages employers to replace existing workers with people from the programmes. It thus tends to push down wages and conditions (Solow, 1998); indeed, Layard (1997), a prominent academic supporter of workfare, sees this as a valuable outcome (see further Chapter 9). Employers are thus also encouraged to pursue low-skill production strategies.

Workfare's impact on labour markets is strongest in localities with high unemployment, thus deepening differences between localities. More generally, the transition from the Beveridge system to workfare has integrally involved **a change in the spatial scale of regulation and its use of spatial difference**. Social insurance was intended to be uniform over the *national* space, congruent with the wider Keynesian project of national economic regulation, consensus and class solidarity. With the rolling out of neoliberal workfarism, the scale of regulation becomes increasingly *localised* and *unevenness* increasingly pronounced. We have seen that the options effectively available to the unemployed vary strongly with local labour market conditions. Correspondingly, the institutional forms of delivery of unemployment benefits have been substantially decentralised (OECD, 1999; on the USA see Peck and Theodore, 2000). The Blair government has given JobCentre Plus offices considerable leeway in adjusting their policies to local conditions; it has claimed that it wishes to encourage local innovation, though programmes remained tightly controlled from the centre (Carmel and Papodopoulos, 2003). In both Britain and the USA, central control increases *political* uniformity by accentuating coercive methods, while failing to overcome differences created by uneven labour markets. In contrast with the post-war policy of moving jobs to workers, pressure is applied to migrate from high to low unemployment areas on pain of losing benefit; in the words of the 1980s minister Norman Tebbit, 'get on yer bike' (cf. section 2.4). In typically neoliberal fashion, this policy ignores social and housing constraints. These geographical aspects of workfare fit with the neoliberal articulation of space: accentuation of uneven development, and encouragement of spatial mobility of factors of production (section 9.1). In these ways, localism reinforces the subordination of the unemployed to the demands of employers.

5.10 CONCLUSION: THE NEOLIBERAL STATE EXACERBATES INEQUALITY

The welfare state, even in its heyday, failed the poor in many respects. It was substantially geared to the demands of employers, and thus often gave the least benefits to those in a weak position in the labour market and to the retired. These problems have been worsened in the last three decades not only by increased unemployment and worsening jobs but also by neoliberal reforms of the state itself. Cuts in spending to fund tax cuts have hit both benefits and public services. Individuals and families are to bear greater responsibility and burdens for their own reproduction, implicitly in the case of service cuts, explicitly in the case of benefit and pension reforms. Increased use of means testing and pressure on the unemployed to take any available work has required stepped-up surveillance of the poor.

The administration of both public services and state incomes has been substantially decentralised, though with central control of funding and service content *intensified*. Both have become more strongly subordinated to local prosperity: funding of public services to the local tax base; service units to the incomes and cultures of their catchment areas and to local business demands; national insurance benefits to local wages; housing benefits to local costs; and outcomes for the unemployed to local labour markets. Inequalities in the welfare state between localities and neighbourhoods have thus deepened (Pinch, 1997, 1999). This consequence is intended: neoliberalism aims to force local people to accept the 'verdict of the markets' on their locality, and thus to compete better within wider spaces (section 9.1).

Neoliberal reforms mark a shift from social to private responsibility for overcoming poverty. They increase the number suffering poverty in incomes, in service availability and service outcomes. People's life chances from cradle to grave are made more directly dependent on labour markets, and the support they can obtain from their family; competition in labour markets and household conflicts are thereby exacerbated. The reforms increase inequalities both between the poor and the better off and among different social groups within the poor.

6

SOCIAL REPRODUCTION
AND SOCIAL EXCLUSION

6.1 WORKING-CLASS REPRODUCTION:
PRIVATISM AND INEQUALITY

Poverty is not solely due to low incomes and inadequate public services, but also arises from the relations of social life characteristic of capitalism. Working-class people reproduce themselves by earning a wage or drawing a state benefit, sharing this income within private households of varying composition, purchasing goods and services, and performing unpaid work for their household or neighbours. In this chapter we examine these processes and their spatiality, and how they deepen poverty.

In the very long term, the form of these relations has changed in a number of connected ways (cf. section 1.1):

- as the productivity of labour increases, the volume of commodities consumed increases and their particular social uses and meanings become more important;
- the purchase of commodities replaces some domestic labour, but in other ways adds to it;
- the family household tends to become less dependent on neighbours, but more dependent on state services;
- the increased dependence on commodities intensifies the need for wage income, and married women in particular are drawn increasingly into wage work. Women's work is thus partially transferred from the home to the capitalist workplace;
- the mass production of consumer commodities is associated with the centralisation of ownership in the industries producing them and with the concentration of consumer services into fewer, larger sites.

For business, these dynamics expand sales and the size of the labour force, and thus augment the accumulation of capital. For working-class people, however, the effects are more mixed. Average living standards do rise, but women's dual roles tend to become more onerous and conflictual. The individual or family becomes more socially isolated as neighbourhood ties erode. Low income has more serious consequences than before. Commodity consumption for the poor is of low, and in some respects declining, quality, and the poor are particularly vulnerable to the spatial strategies of service corporations. Thus, inequalities in living standards tend to become more significant, including those in housing and use of space. In the rest of this chapter we examine these changes and the social relations of reproduction within capitalism on which they rest. We are concerned with changes over the last century, in particular under neoliberalism.

6.2 THE RISE OF THE NUCLEAR FAMILY AND CHANGE IN GENDER DIVISIONS OF LABOUR

Over the last hundred years or so, the amount of work of directly caring for others has been reduced principally by free or subsidised state services – schools, nurseries, health services, care homes. Other types of domestic work have been lightened by the use of durables and by buy-in services (Coontz, 1988). But increased consumption of commodities and public services creates new demands on time – shopping, repairs, trips to services, taking dependents to school or hospital. Since the 1950s the increasing sexualisation of everyday life has required increasing work on and around the body. For these reasons, the total amount of domestic work has not significantly decreased over the last century, and occupies more time than waged employment, particularly for the poor. Reproduction work still falls predominantly on women; poorer women with dependents thus suffer onerous 'triple working days' (Little *et al.*, 1988).

Over the same period, the extended family in the working class has tended to shrink to a nuclear family of parents and their biological children. Extended families comprised large households but also numerous relatives within a few minutes' walk, presided over by their matriarchs. Mothers could call on their female relatives to help with childcare. Children were often brought up for years by relatives when their parents couldn't cope (an arrangement now bureaucratised and hindered by the state). Older women were often visited by their daughters daily (Young and Willmott, 1957). The shift to the largely autonomous nuclear family has been underpinned by more affordable housing enabling independent living, lightening of some domestic tasks, the state pension, smaller numbers of children per parent, and by greater propensity to move to obtain jobs. The effect, however, has been to reduce support from relatives for domestic work. Working-class carers are now more isolated and have greater difficulty in taking up employment. The elderly, people with disabilities, and some children receive less care from their relatives.

In the late nineteenth and early twentieth century better-paid working-class men were able to obtain a 'family wage' that enabled their wives to avoid employment. Since the 1960s, however, women with male partners have increasingly taken up waged employment. This has been partly because of reductions in some kinds of domestic work. But, since the 1970s, it has also been under pressure of deteriorating male employment conditions for men (section 4.4): their female partners have increasingly gone out to work so that the household can achieve consumption norms. In consequence, the two-wage couple has become the norm. This development has the *potential* for greater gender equality; but the reality has largely been, as already noted, for women to continue to do the lion's share of domestic work, so that the net result is an intensification of their working day. Moreover, the new social *norm* of two wages means that working-class couples with one or no wage are much poorer relative to the average household than trends in individual wage distribution would suggest. Indeed, under neoliberalism the number of such households, particularly those with no wage, has increased sharply, due to rising (male) unemployment and to (women's) increasing caring responsibilities (Sklar, 1995a). Household income increases from dual wages have mainly benefited the better off, thus increasing class inequality (Crompton, 1997; Leira, 2002).

The lightening of some domestic work, the expansion of jobs denoted 'female', the decline of well-paid male employment and, since the 1970s, the influence of feminist ideas, have meant a sharp increase in the number of women living outside of the heterosexual couple. Many of these were formerly in a couple, and many of these are single by their choice: women currently petition for three-quarters of divorces. In particular, in the last 30 years singledom has increased sharply for mothers, most of whom, contrary to popular prejudice, were formerly in a relationship with the father: 20 per cent of children are now in lone-mother families. But again, while this development has been underpinned by some material improvements for women, it has resulted in high levels of deprivation for poorer lone mothers. Jobs available to them are of very poor quality. Given low wages, childcare is an enormous problem: nursery places and after-school care are prohibitively expensive, childminders are rare in poor localities because their housing does not meet minimum standards (Speak and Graham, 2000), and state provided childcare is rarely available. Thus a combination of change in gender roles (the possibility of singledom) and *lack* of change in them (women's wages, childcare) have produced widespread deprivation for single women.

6.3 THE DECLINE OF NEIGHBOURHOOD SURVIVAL STRATEGIES OF THE POOR

The poor have traditionally survived through the material and moral support of their neighbourhood, which enabled them to maintain some sense of agency, control and dignity. Caring and housework was not purely private but shared with extended family *and* non-kin neighbours. They supplied not only help with domestic work but constituted mini-welfare systems that helped to solve problems in family relations, finance and housing. Services such as childcare, repairs or haircuts were available on a reciprocal basis (Pahl and Wallace, 1985). There were common community rhythms – the Monday wash, the men in the pub after work (Hey, 1986). What they did not provide was privacy or room for the abnormal (Hoggart, 1957).

Kin and neighbourhood also provided access to jobs, particularly for men. Better jobs in unionised workplaces, especially, required someone to 'speak for' you (McKibbin, 1998: 120). Moreover, communities facilitated 'ducking and diving' strategies – juggling short-term credit, using the pawnshop, doing a moonlight flit, working in informal jobs – the entrepreneurship of the poor (Ross, 1983; Tebbutt, 1983). Cheap illegally acquired goods were available – durables, food, clothing, alcohol and cigarettes. Many neighbourhoods had small illegal shops run by older women in their homes. These means of survival were legitimated by the local culture, which consequently tended to be hostile to policing of the neighbourhood. Significantly, Oscar Lewis's (1979) well-known theory of 'the culture of poverty', while it portrayed a 'deviant' culture, argued that this was a rational way of coping with poverty.

This system has, however, been eroded. Indeed, erosion has been *en train* since the beginning of industrialism. Workers brought with them pre-industrial practices which helped them to bear the new conditions: to fish and poach game, keep animals, grow food, collect their water from the roof, do independent artisanal work at home, have lodgers, be given or take goods from the workplace, and use the streets for leisure. But in the nineteenth century these rights were steadily eroded: the keeping of animals and the use of water butts were made illegal; home work and lodgers were prohibited in charitable and municipal housing; street games were repressed. Subsistence thus became increasingly dependent on commodities so that regular wage work became more important (sections 1.1, 1.4 and 1.5).

During the twentieth century networks in poor neighbourhoods were weakened by greater household time spent in waged work, increasing commoditisation, growing state services and regulation, and suburbanisation. All of these were associated with increases in living standards for the majority, and, to some extent, for the poor. Yet their paradoxical effect has been to weaken the *social structures* through which the poor survived, and to reduce their control over their lives and ability to recover from setbacks. Thus, women's increased involvement in waged work has reduced the time and energy available for helping family and neighbours; circularly, the declining availability of this help, and consequent greater need for commodities, has been one of the reasons for increased employment. Greater dependence on commodities has eroded practical skills for work and creativity in the home. Expanding public services have been designed to individualise their users rather than develop community cooperation (section 2.3; Chapter 5).

Slum clearance dispersed longstanding communities by failing to re-house them *in situ*. Indeed, this was often an aim of planners (Burns, 1963; Davies, 1972) who sought to break up what they saw as a negative culture of poverty (section 2.6). Council housing opportunities for young adults often located them far from their neighbourhood support networks (Speak and Graham, 2000). Movement to better-quality housing was promoted as a way of improving social life through the privatism of the nuclear family (section 2.5). This was all too successful: social networks were disrupted, and, though they were sometimes recreated in the new neighbourhoods, they were weaker than before (Young and Willmott, 1957). Moreover, the older settlements were leached of those inhabitants with the stronger employment records.

These changes, together with the shrinkage of the family, have had the effect of enclosing the poor household or individual (Li *et al.*, 2003). Help with caring work, moral support in crises and job networking have been weakened. Many of the poor consequently suffer from isolation, sometimes reinforced by depression and low self-esteem. Neoliberalism has worsened this fragmentation by accentuating competitiveness and individualism, and by removing or weakening the collective institutions – shared employment, unions, welfare services – which nurtured trust, processes which have affected the poor most adversely (Foley and Edwards, 1998). Moreover, weakening of neighbourly ties has made it more difficult to deal with bad neighbours. It has reduced the collective pressure and actions that kept noise, rubbish dumping, public violence and vandalism in limits, so that these have become major problems of poor areas.

This is not to argue that mutuality and community solidarity have disappeared among the poor. Indeed, community networks may be strongest in the poorest working-class neighbourhoods because of need. Forrest and Kearns (2001) examined the research on four notoriously poor localities and found that they had active social networks and bonds of trust. The strength of community ties varies strongly between neighbourhoods, depending partly on the degree of population stability (see also section 3.4; Cattell and Evans, 1999; Johnston *et al.*, 2000). The diagnosis of complete social fragmentation is an ideological one: it deflects attention from external conditions creating poverty and legitimates the 'capacity building' approach to poverty alleviation (section 8.5). Nevertheless, social-spatial modernisation has eroded many of the traditional survival mechanisms.

6.4 AUTHORITY, CULTURAL CAPITAL AND STIGMATISATION

We saw in Chapter 3 that poverty is persistent over time for individuals, for particular social groups and areas, and between generations. To a considerable extent this is due to what Bourdieu (1986) has called 'cultural capital'. This comprises characteristics such as accent, physical demeanour, social behaviours, cultural tastes, and, in a somewhat wider definition, attitudes to life and work; it is distinct from technical skills. Cultural capital is acquired in childhood, primarily from family and neighbourhood (though also from the mass media), and is therefore class specific. It is the set of signs through which each class identifies and reproduces itself. Selection into jobs is strongly mediated by cultural capital because it influences formal success in education and because of discrimination by employers. Moreover, access to social and political networks depends on the status and authority given by the individual's cultural capital. Cultural capital is the basis for instinctive *solidarities* within both the upper class and the middle class that exclude others. Higher-class cultural capital gives those possessing it a rational basis for expecting esteem, lending a self-confidence, which further reinforces their social advantage; conversely, the poor often *expect* to fail. Higher classes therefore go to great lengths to have the appropriate traits instilled in their children.

Working-class cultural capital is specific to particular neighbourhoods and localities, and is *recognised and validated* only in those spaces. In contrast, higher-class culture is regarded as 'universal' and recognised in wider spaces. Opportunities for poorer people are thus limited because their culture is not portable: cultural exclusion is strongly spatial. Moreover, lower-class cultural capital is durable, and tends to resist change through formal education (section 5.3). For particular social groups, generations may be required to consolidate effective higher-class culture. Before the growth of the African-American middle class with affirmative action in the 1960s, nearly two-thirds of the sons of black professional fathers ended up as manual workers, as their class culture was insufficiently strong to counter racism (Kerbo, 1983).

Despite the mass media, the class gap in cultural capital is probably not narrowing. Deskilling of manual and clerical jobs (section 4.4) has eroded the culture of the skilled working class. On the other hand, even low skill jobs serving the public now typically incorporate an element of performance, and employers usually require this performance to be within middle- or high-income culture. An 'economy of signs' (Lash and Urry, 1987) tends to widen differences dependent on ability to interpret (and design) those signs. Under neoliberalism, widening income differences have led to diverging body practices (nutrition, exercise, health, clothes), to sharper differences between neighbour-hoods (section 6.6 below), increased use by the middle class of private schools and tuition, and sharper snobbery (section 7.2). Thus everyday behaviours still, perhaps increasingly, exclude the poor.

6.5 THE MARKET VERSUS DECENT HOUSING FOR THE POOR

As the largest item of everyday expenditure, housing manifests in the sharpest way the problems of commodity consumption for the poor. Because of these problems, during the twentieth century the market was mediated, but by no means tamed, by state intervention. Until the First World War, housing for all classes was mainly privately rented. For working-class tenants rents were high and quality was poor or abysmal. From the 1920s, working-class campaigns resulted in **state provision**; in its heyday in the 1960s it constituted a third of the housing stock. Council housing softened the pressure on people to take any job on offer, strengthening their bargaining power in the labour market. However, it largely excluded the poorest and black people, who remained dependent on private renting or were offered the very worst council housing (Rex and Moore, 1967). Moreover, during the 1960s the quality of council house building started to deteriorate, and from the 1970s its maintenance also. In the 1950s and 1960s **private renting** was regulated to improve security and quality and moderate rents; but, in a typical revenge of the market, this accelerated the decline of the sector. Thus, even in the post-war boom, state regulation failed to provide decent housing for the poorest (Kincaid *et al.*, 1962).

Neoliberalism, however, has made things worse. The Thatcher government claimed that 'freeing' markets in housing would enable supply to meet demand, even from the poor (section 9.1). **Private rented** housing was deregulated; but this resulted not in expanding supply but in worsening quality and greatly increased rents. Neoliberalism has set out to eliminate the **council housing** sector (Lowe *et al.*, 1991; Cole and Furbey, 1994). Council tenants were given the right to buy at below-market price; local authorities have been prevented from constructing new housing or even borrowing to repair their stock, and have been compelled to increase rents. Sales to occupiers have leached out the best quality council housing and better-off tenants, so that council housing has become a residual tenure for the poorest. By 1993 two-thirds of heads of households with council tenancies were unemployed (Rahman *et al.*, 2001: 82). Paradoxically, in growth regions access to social housing has become difficult even for the most needy (Butler, 1998).

The Blair government has sought to deliver the *coup de grâce* to council housing by transferring it to Third Sector housing trusts; tenants vote whether to transfer, but only a 'yes' vote releases the funding for repairs. The housing is then outside democratic state control, rents have to rise to market levels, and housing associations are quicker to evict tenants. On this basis, some council tenants have defied the blackmail and voted against transfer. But the poor, concentrated into state and Third Sector housing, are now forced to spend a higher percentage of their income on housing than did previous generations.

From the 1920s **owner occupation** grew with substantial state subsidies. Since the 1970s the decline of private renting and the residualisation of social housing have made owner occupation the social norm; under the Thatcher governments it increased from a half to two-thirds of dwellings. This affects the poor in two ways. First, a significant minority of the poor are owner-occupiers. Already in the 1960s many of those excluded from council housing – the poorest and black and minority ethnic groups – chose to buy rather than pay high rents for slum housing, resulting in them living in their own low-quality and over-crowded accommodation. Working-class people who bought houses when in employment are often unable to afford to maintain them when in retirement, sometimes because the houses are under-occupied. One in seven mortgage payers is in part-time work or has no work (Palmer *et al.*, 2004: 108). They are vulnerable to interest rate rises and, since the de-regulation of mortgages in the 1980s, unreliable financial instruments such as endowment mortgages.

Fully one half of unfit housing is owner occupied. Thus, the *achievement* of owner occupation by poorer people produces new problems.

Second, much of the working class is now excluded by cost from the purchase of a home. The dominance of owner occupation has not produced a supply sufficient to keep prices near to building costs, as free market theory suggests it should; on the contrary, since the 1980s house prices have increased much faster than average incomes and the price of new housing incorporates an enormous surplus profit (partly expressed in land price). This premium is especially high in the 'prosperous' regions, but is found even in the poorest. The central reason is that commercial house builders have reacted to strong prices not by increasing supply but precisely by restricting it, in order to sit on appreciating land assets: the rate of house completions of all tenures in the last five years has been around a third of that in the 1950s. This has been exacerbated by lack of state infrastructural investment for new residential areas, and by ever-worsening inter-regional uneven development in employment and hence demand: capitalist market mechanisms poorly coordinate labour markets with housing markets (Allen and Hamnett, 1992; Hamnett, 1998). Rent rises in social housing, and acute shortages of it in the growth regions, have further pushed up freehold prices.

The resulting house price inflation has priced middle-income, let alone poor, new households out of the market: in London only 22 per cent of new households can afford a mortgage. The lower-paid cannot afford to move to the regions where jobs are available; indeed, new households are having to move out of these regions. Ability to buy depends increasingly on the lottery of inheritance. Even the average wage earner is suffering from falling space standards, decentralised locations and increased commuting (Thomas and Dorling, 2004). Thus the problems of the poor are here in a continuum with those of the majority. Finally, house price inflation has greatly widened class inequality in assets, albeit with the complexities of location and family inheritance (Allen *et al.*, 1997).

The neoliberal prescription has thus worsened access to and conditions of housing of the poor across all tenures. But this failure is inscribed in the nature of housing as a commodity; and state intervention, even during its post-war heyday, has been insufficiently strong to bend the market to the needs of the poor.

6.6 RESIDENTIAL EXCLUSION AND CONCENTRATION OF THE POOR

Capitalist towns and cities separate different income groups by area. The market in housing, whether rented or owner occupied, combined with people's wish to live with (at least) their 'social equals' and in the best possible physical environment, tend to create *single-income* neighbourhoods and districts. This is a dynamic process: when a household's incomes rises it tends to move to a

Figures 6.1 (a and b) Houses being cleared for gentrification. In 2001–4 Newcastle City Council evicted hundreds of residents in Scotswood from structurally sound houses to make a site sufficiently large and close to the river Tyne to attract a developer to build an enclosed middle-income estate. This was presented as a 'new' approach to 'regenerating' Scotswood in the light of 'failures' of previous urban programmes. It was part of a wider strategy to shift from a proletarian to a post-industrial city (section 10.4) (source: (a) Duncan Fuller (b) Jamie Gough).

'better' area, producing successions of newly desirable areas on the one hand and neighbourhoods abandoned to investment on the other. Thus, since the eighteenth century, the poor have been largely confined within their own neighbourhoods. Segregation has been moderated, however, by the sometimes slow pace of movement of the better off, by poor people living at high densities in high rent accommodation, and by older, good-quality housing having a complex spatial distribution.

In British and (somewhat later) US cities the pattern has been of outward movement of the better off to newly-built suburbs: the rich from the eighteenth century, the middle class from the early nineteenth century, the better-off working class from the late nineteenth century through to the 1950s. Social democratic reformers from the Garden City movement through to post-Second World War planners saw the leafy suburb as a solution not just to the housing problems of the poor but to poverty in all its forms (section 2.5). But the poor could not afford the housing costs of the suburbs nor the costs and time of commuting, and thus remained in their old areas. In consequence, by the 1960s inner city areas were largely poor.

State intervention has actually reinforced these divisions. In both Britain and the USA suburban expansion has received state subsidy (infrastructures, tax relief on mortgages) (Walker, 1981). Land use planning has set maximum densities and restrictions on conversion, ensuring that the neighbourhood cannot be cheapened by builders. In Britain, as elsewhere in Western Europe, public housing has been built mostly in large estates sited on cheap land, either in old working-class areas or (more common on the continent) on peripheral green field sites. In Britain this was effected partly by the refusal of better-off suburban authorities to build council housing, and partly by central government funding regimes favouring high rise as experiments in the industrialisation of building. The majority of public housing has thus been in distinct – thus lower income – neighbourhoods, *mirroring* market patterning. The residualisation of council housing since the 1980s has then automatically led to the creation of (increasingly) poor *neighbourhoods*, exacerbated by the disproportionate sale of council housing located in higher income areas. The outcome of these market and state processes is strong spatial concentration of the poor in contemporary Britain (section 3.6; Wadhams, 2002).

Spatial concentration of the poor is deepened for **black and minority ethnic groups** by racism. The extreme 'racial' segregation in the USA is well known; many African-Americans do not see white people in any capacity. This arises from a mixture of income segregation, need to be near central-city secondary jobs, and racist exclusion from many suburbs (McKay, 1976: 134). In Britain, from the 1950s black people were excluded by racism from most areas of whatever tenure. They were able to rent or buy only the worst houses; these areas then became easy for black people to move into, cumulatively reinforcing ghettoisation (Rex and Moore, 1967). These patterns have continued to the present day. Ironically, as social housing has become residualised, BME groups have achieved better access to it. But continuing racism in its allocation, and

and self-policing private shopping malls, private car parking lots, and a range of privately owned places of leisure and sport . . . the central iconic building . . . is the shopping mall itself.

(Taylor, 1999: 60)

These are tightly policed. 'The Disneyfied city centre shopping malls become no-go areas for those classified as not belonging, by virtue of their credit ratings or their ethnic background' (Lyon, 2001: 67). CCTV discourages the discordant presence of the poor (McCahill and Norris, 2003). 'Preventative surveillance' is based on typification of 'whole groups, populations and environments' (Cohen, 1985: 127). The new spaces are designed to discourage passing the time of day and hanging around, and they do not allow any political activity. Nothing should interfere with the pleasure of being with other prosperous people, nor jar the atmosphere of affluence, the dreamy immersion in luxury. Shopping, leisure and tourism are thus promoted by creating a safe bubble within 'a fortified city' (Judd, 1999). These spaces represent the triumph of consumer service corporations over socially-inclusive public space.

6.8 IMMOBILISED BY THE CAR

The socially excluded have less mobility than the better off. The proximate cause is their lack of access to cars and dependence on public transport. This is partly due to lack of buying power: low-income households are able to spend only a quarter as much on transport as households on average incomes (Palmer et al., 2004: 100). Thus only 45 per cent of the economically inactive and unskilled manual workers are in a household with a car, and only 18 per cent of those over 65 living alone have a car. There are further barriers to car use: children and people with certain disabilities have to be driven; women in heterosexual couples with one car usually do not have the use of it during their partner's working hours. Compared to car use, public transport, particularly the bus, is usually slower, less predictable in its timing, less comfortable, and less convenient. The regulation of buses initiated in the 1920s was all but abolished in the 1980s; the predictable result has been sharp increases in fares, dilution or withdrawal of services, and increased unreliability (Dobbs et al., 2004). Moreover, the state now lacks power to plan public transport in relation to land use. Thus a new industrial estate designed by Newcastle council to provide much-needed jobs for residents in the West End of the city has no bus service: deregulation makes planning for the poor impossible.

Lack of mobility of the socially excluded is particularly severe in rural areas. Nearly a third of rural settlements have no public transport at all and 75 per cent are without a daily bus service. Yet rural residents have to travel on average 40 per cent further each week than those in urban areas (Simmons, 1997). The paucity of these services is also one factor in deterring poor people in cities from visiting the countryside for leisure.

The poor's disadvantage here might be regarded simply as an effect of the supercession of an inferior by a superior transport technology, and the poor's temporary inability to afford the latter. But, at a deeper level of social choice, disadvantage has been caused by the logic of capitalist transport industries combined with patterns of land use driven by private profit. First, capitalist consumption is centred on the private household. Thus (a) for capital, car production and sale is the natural transport form; (b) it is assumed that car users have a right to use roads untrammelled by considerations of congestion, ecology or road construction cost; (c) cars are purchased and used irrespective of the effect on collective mobility, so that decreased public transport use leads to declining quality in a well-known vicious circle; (d) cars, *qua* individual purchases, become linked to status: poor youths who take a car without the owner's consent are seeking to use a commodity with high status as well as high speed; conversely, bus and most rail transport now has low status.

Second, commercially viable, and efficient, public transport needs coordination across the city-region; this means cross-subsidy between routes and planning them, which go against the capitalist logic of their *individual* profitability. Third, private investors in employment, consumer services, commercial space and housing do not have to coordinate their sites with public transport. As car-owning workers and consumers are the important ones for investors, sites become increasingly dispersed and journeys longer. In 1952 people travelled on average 86 km per week; 50 years later this was 251 km (Department for Transport, 2003). Dispersion of sites then makes jobs, services and residential areas harder to serve by public transport in a further vicious circle.

The poor's lack of mobility thus derives from capitalist production dynamics in transport and the built environment, including the lack of spatial coordination in investment, and from the dominance of private-household consumption. The state potentially can improve accessibility through regulation of transport and land use; but it has to do so in the face of these fundamental capitalist relations.

6.9 DAMAGED BY COMMODITIES

Long-term growth in the volume and types of commodities consumed puts increasing pressure on the poor to consume them. First, new products become *necessities*: the car (as just discussed), the TV (to be in touch with the contemporary world), the mobile (for young people to have a social life), and so on (section 3.1). Culturally-loaded commodities become important for the formation of ersatz identities ('fcuk for England' T-shirts, for example). Second, when commodities which provide a certain service become the norm, they often cause the loss of skills to produce that service: thus in the twentieth century skills such as cooking and making music have been severely eroded. Alternatively, home activity and the development of skill come to depend on purchased goods. In either fashion, the commodities become essential.

communities have weakened (section 6.3). Moreover, Walklate and Evans argue that criminals are now more opportunistic and therefore harder for the community to bring into line.

Thus the rise of petty crime and disruptive behaviour by poor men, in particular in their own neighbourhoods, can be traced to the long-term commoditisation of daily life and to economic and cultural impacts of neoliberalism.

6.11 THE IMPACT OF CRIME ON THE POOR: FEAR AND POLICING

Crime in poor neighbourhoods impacts onto all their residents and, in a different way, onto those who are criminalised by the justice system. **Residents** suffer from higher levels of property crime than in better-off neighbourhoods (even though the pickings for robbers are richer in the latter), and also from higher levels of violent crime. Violence accompanying robbery or rape, or carried out for its own sake, disproportionately affects women, BME people, the young and the old (Mirrlees-Black *et al.*, cited in Abercrombie *et al.*, 2000: 528–9). Mediated by perceptions, this violence strongly affects the spatial life of these groups. Women choose carefully when and where they walk; many of the old do not go out after dark, for fear for both themselves and their homes; many poorer Asian women stay very close to home and do not go out to work; and many children fear walking to and from school (Pain, 1995; 1997). The fear of violence thus imposes severe constraints on the social and economic life of most of the poor, with the partial exception of young and middle-aged men.

It is not only low-income victims of crime who suffer from it, but poor **criminals and those thought likely by the justice system to be criminals.** First, blue collar crime is far more policed, investigated and prosecuted than white-collar and corporate crime, and is more likely to lead to prison: the justice system has a systematic bias of class, in both the normal and Marxist senses of the term. One measure of this is that in popular discourse, the media and academia, 'crime' nearly always *means* blue collar crime. The methods used by the justice system are predicated on the fact that most of its subjects are working class: violence and frame-ups by the police; bias by judges in directing juries and in sentencing; and the appalling conditions in prisons. Regarding the latter, British prisons are now severely overcrowded, allow prisoners out of their cells very little, offer few opportunities to acquire new skills, and allow systematic violence of some prisoners against others (whites against blacks, for instance). The *majority* of prisoners have mental health problems, partly as a result of this treatment, but get very little help. The conditions in women's and youth prisons are if anything worse. This class bias in the justice system is an extension of the disciplinary nature of the capital–labour relation (sections 1.3 and 1.4). The exclusion of the public from the spaces of the justice system (police station, court, prison), despite their nominal democratic control, makes it easier for these abuses to continue.

Second, *within* the working class the justice system targets particular groups. Black people in Europe and the USA are more likely than whites to be investigated, subjected to police violence, and prosecuted (section 3.5). In 1995 7 per cent of all African American men were incarcerated and one third of all young black men were under some form of state supervision (Western and Beckett, 1999: 1035); Louis Wacquant has argued that the justice system is *the major* structuring institution of these men's lives. Being a 'useful' worker makes a difference too: 'The police may be more likely to divert an employed offender from prosecution seeing them as less likely to continue their offending and, similarly, courts may be deterred from sending an employed person to prison' (Croall, 1998: 114–15). These biases continue the elite's longstanding fear of the 'underclass' (section 1.5).

In the last fifteen years or so, the justice system has increasingly targeted **disorder** in public spaces. New York City's mayor Rudolph Giuliani famously and influentially introduced 'zero tolerance policing', aimed principally at purging commercial spaces of discrepant users: the homeless, beggars, bohemians, politicos (cf. section 6.7; N. Smith, 1993). The current British government has introduced Anti-Social Behaviour Orders (Asbos) to be used in residential areas. These are, in part, a response to demands from residents of (mainly) poor neighbourhoods arising from the declining ability of communities to impose considerate behaviour (sections 6.3 and 6.10). But the dependence on policing has produced some grotesque outcomes: interdicted behaviour has included sarcasm, playing football in the street, not riding a bicycle, and wearing head-gear in public. Orders can severely restrict movement. A third of Asbos are breached, which can attract a sentence of up to five years' prison. The prejudices of the state concerning the 'underclass' make policing of good behaviour problematic in a way that community action was not.

6.12 SOCIAL LIFE DEEPENS POVERTY

In this chapter we have seen many problems in the social life of working people under capitalism, which play a major role in exacerbating poverty and the experience of poverty. We have seen that the articulation of spaces is central to these problems: the relations between the spaces of residence and waged work, between private home and public neighbourhood, between neighbourhood and the spaces of local services. We have analysed these problems *at different levels of historical abstraction*. We can summarise our argument by drawing the threads of these together, thus highlighting how one is expressed in the next.

First, social life **in capitalism in general** generates poverty:

- The individual and family as the fundamental units of reproduction – income sharing, commodity purchase, domestic work – privatises poverty and the struggle to deal with it.

7

EXCLUDING SOCIETIES

7.1 EXCLUDED IN EVERY FIELD OF LIFE

In the previous three chapters we have examined, respectively, the economic, state and social underpinnings of poverty. We have seen many connections between each of these. Some of these connections constitute vicious circles, where (variously) poor jobs and unemployment, administration of state benefits, poor treatment by public and private service providers, consumption of low quality housing and other commodities, and social relations in poor areas compound each other. These can be vicious circles for an individual or a household: sudden adversity – loss of a job, a cut in benefit, ill health, a major house repair – can set off a spiral into poverty which then becomes very hard to escape.

In this chapter we examine three aspects of the construction of poverty that combine inextricably the economy, the state and social life. First, we examine hostile attitudes to the poor and oppressed, arguing that these are reproduced by, and reinforce, material causes of poverty. Second, we discuss how different forms of social oppression can lead to poverty, and how, reciprocally, poverty deepens social disadvantage and prejudice. Finally, we suggest some key aspects of economy, state, social life and culture that differentiate national societies in the way that they generate poverty and exclusion.

7.2 POPULAR ATTITUDES TO THE POOR

We have seen repeatedly the often-aggressive attitudes of the elite and gate-keepers to the poor. But ordinary people, too, are often hostile to or dismissive of the poor. Indeed, sections of the poor are often hostile to others; and the poor may internalise this hostility as self-blame. This not only makes daily life harder to cope with, but also blocks collective organisation against poverty; it is thus an integral part of the relations of exclusion. Hostile attitudes are partly created by state policies and ideologies and propagated through the mass media (Chapters 1 and 2). Differences in culture, 'cultural capital' and their geography (section 6.4) are another source of vilification. But attitudes are also shaped by

128

Figure 7.1 Landscape in Elswick, Tyneside, 2005 (source: Jamie Gough).

the (spatial) problems of daily life of both the poor and the better off. We may distinguish three types of process.

First, the poor often experience hostility arising from **competition for resources** such as jobs, state benefits, housing and living space, and welfare services. This competition is conducted in and through space, and indeed control of space is sometimes its aim (Cox, 1989). In this perspective, there is nothing *morally* wrong with the poor, but they are a material threat. Thus those towards the bottom of the labour market may see themselves being under-cut by people prepared to accept even worse wages and conditions. This idea is usually strongest where distinct social groups compete: indigenous or white workers feeling they are being undermined by immigrants or blacks, men feeling their jobs are being devalued by the employment of women (Cockburn, 1983), or, as is now common, older workers made to compete with school leavers on inferior wages and conditions (see section 4.4). Competition for social control of residential space may also engender hostility. The non-poor often fear the encroachment of the poor because of loss of status and anticipated rise in crime, and the poor often similarly resent particular groups among their neighbours. Both may seek to exclude black people from their residential space (section 6.6). The better off may feel that the poor spoil public or private services, which they use (sections 5.3 and 6.7).

Sections of the poor and the nearly-poor may feel that they are in competition with others for state benefits, and for public services targeted on the poor such as social housing; these may be perceived as a fixed resource for which different groups are in competition. This gives rise to hostility to groups seen as benefit scroungers (the lazy, recent immigrants), and those who are seen as having jumped the housing queue (asylum seekers, young women 'who have deliberately got pregnant'). The government has encouraged this kind of hostility by asking people to report supposed benefit or welfare cheats. Finally, poor men may feel that they are in increasingly sharp competition with women not only in the labour market but also in the sharing out of domestic work within households, and express this in personal and general sexism.

While these forms of hostility are materially based, they are often founded on false theorisations; indeed, they may contradict each other: asylum seekers are here to take 'our' jobs *and* to sponge off the welfare state. All these forms of competition are most strongly experienced by those who are nearly poor, and by sections of the poor against others: Golding and Middleton (1982: 169–72) found that the greatest hostility to social security claimants is from those in low paid employment. However, some forms of competition, such as in residential space and services, are experienced by the better off and shape their attitudes.

A second, very different source of hostility is **fear of what the poor advertise**: the possibility of falling down the social ladder. This fear is endemic in contemporary society characterised by sharp economic, social and cultural insecurity (Vail *et al.*,1999; Beck, 1992). Blaming sections of the poor and excluded for this insecurity is irrational transference, but nonetheless often strongly felt. Space is here *imagined* space, contrasting a putative good, safe space of the individual with the threatening territories of the poor. In contrast to material competition, fear of social decline produces hostility in the mode of moral outrage – 'they should pull themselves together and try harder, as I do'. Thus those who have been harmed by capitalist modernity may cling onto a supposedly stable, familiar identity, identifying modernity with the despised Other. This may be a national or ethnic identity, in which minority ethnicities are seen as a cultural threat (Betz, 2002). Or men may cling to a traditional masculine identity in which certain types of women or behaviour by women are seen as undermining their social role (section 6.2). The contemporary experience of work for the majority of the population, with its demands for ever-greater effort (section 4.4), can cause resentment against those seen as lazy, work-shy or unproductive. The individual's work effort is morally validated and thus made tolerable by contrasting it with the feckless Other. Flecker (2002) has suggested this as a source of contemporary xenophobia in Western Europe among workers in higher status jobs and prosperous regions.

Third, hostility to the poor may be based in **sub-conscious associations of the poor** with dirt, disease, danger, or forbidden eroticism (Sibley, 1995). These feelings are constructed out of deep, perhaps trans-historical psychological structures of sense of selfhood, fear of encroachment, sexuality and so on (O'Connor, 1987). The imagined space of the body is central here. Such hostile fantasies, however, need the support of *historically specific* social relations to be mobilised and are directed at *socially defined* 'inferior' groups.

Social practice may produce not only hostility but also **indifference and anomie** towards the poor. These are fostered by the social-spatial separation of the poor from the better off in employment, residence, services, transport, and friendship networks (sections 4.2, 6.6, 6.7 and 6.8). They are reinforced by a lack of understanding of the real economic and social ties between the poor and better off, and the 'culture of mutual indifference' (Geras, 1998) which is prevalent in neoliberalism.

Finally, **the poor may blame themselves for their situation**, resulting in low self-esteem (Handler and Hasenfeld, 1997). This sometimes has 'visible' consequences: voluntary isolation, depression and mental illness (Coffield *et al.*, 1986) or addiction to hard drugs (the latter epidemic among youth in the former pit villages, for instance). In an individualistic society, it is natural to blame social failure on oneself (Lembcke, 1993).

Neoliberalism has tended to worsen all these forms of (self-) hostility through its promotion of individualism and competition and its creation of material insecurity. The mainstream culture of neoliberalism deepens the problem. A prominent trope is anger and aggression to express the *immediate* frustrations of the *individual* – as distinct from angry collective expressions of well-considered hurt. This trope is strongly represented in action films, 'reality' TV, talk-back radio, and electronic games. It tends to result in unthought-out aggression towards the excluded, and to fruitless, private expressions of anger by the excluded.

All these 'structures of feeling', however ill-founded, contribute to disempowering the poor. Most immediately, the daily experience of hostility rubs salt in the wounds of their material situation and undermines the energy and self-respect needed to cope with it. Hostility also weakens political organisation to ameliorate poverty. Self-blame tends to inhibit organising against poverty and making demands on society. The hostility of sections of the poor towards others construed as competitors inhibits their collective organisation. The hostility or indifference of the better off inhibits solidarity with the poor. In another vicious circle, the place of poverty in society generates attitudes, which undermine collective organisation to address it.

7.3 SOCIAL OPPRESSIONS, CLASS AND SOCIAL EXCLUSION

We saw in section 3.5 that socially oppressed groups are disproportionately in secondary jobs and on low incomes and have inferior access to both public and private services. Moreover, poor people find it harder to 'buy their way out' of social oppressions. The resulting combinations of social oppression and poverty create particularly virulent forms of exclusion. But what are the deeper social-spatial processes, which create these relations between oppression, class and poverty? The answers are enormously complicated (Meiksins Wood, 2002; Callinicos, 2000; Saul, 2002). But in the last three chapters we have seen elements of possible analyses. We can now, very briefly, bring these economic, state, social and cultural threads together.

Carers and women

Women's susceptibility to deprivation centres on vicious circles linking their roles in domestic and caring work with those in waged work. Women's imputed role in the home, sanctified by tradition and reproduced through social conditioning and pressure from men, severely constrains the employment they can obtain. Local industries often use these constraints on women to employ them in low wage and insecure jobs. Women's inferior wages and prospects for promotion then reinforce the tradition of assigning the major share of domestic work to them. The gender division of labour in turn incubates patronising and sometimes violent relations of men to women in the home (Dobash and Dobash, 1983). These vicious circles are more intense the lower the couple's income: the more domestic work there is, the greater pressure there is for a clear division of labour in it. Thus more restricted are the choices for women in their wage work, and a more onerous and constrained triple working day results (Glendinning and Millar, 1992). These vicious circles for poorer women are are thus predicated on both capitalist exploitation and male power, connect them, *and reproduce each of them* (Kuhn and Wolpe, 1978; Barrett, 1980; Hennessy and Ingraham, 1997). The gendering of waged work and the social sphere then impact onto *all* working-class women whether or not they have major caring responsibilities.

The key geographical processes here are the *interdependence* and *conflict* between the spaces of the home and of waged work, combined with the *dominance* of the former space by men and the latter by capital. This geography is mediated by urban layout which obstructs carers' triple working day (Little *et al.*, 1988). *Geographical variation* in gender arises from interlinked variations in the gendering of wage work and gender relations in the home (McDowell and Massey, 1984; Duncan, 1994). The state in some ways reinforces these vicious circles, through the content of education and family services, through poor provision of childcare and public transport, and through the benefit system (see Chapter 5). The ideology of women's 'soft' role in the home and disadvantage in 'public' waged work is then extended into men's domineering behaviour in all public spaces.

Lesbians and gay men

The *immediate form* of oppression of lesbians and gay men is pervasive prejudice. This is partly reproduced by religions, moral-political movements, and the state. But, more importantly, it is both created and eroded by the material-social relations of everyday life. To the extent that powerful economic and social pressures construct the heterosexual family, same-sex sexuality appears as unnatural. But this ideology has been weakened in the very long term by the material erosion of the heterosexual family and associated ideologies of gender difference (section 6.2). Moreover, the ability to lead a lesbian or gay life is materially dependent on the ability to reproduce oneself outside the family, and this has grown in the long term through increased consumption of commodities and public services (Gough and Macnair, 1985; Greenberg, 1988).

By the same token, however, it is harder for the poor to escape the heterosexual family, whether this be a lesbian who wants to escape her marriage or a young person seeking independence from his or her parents. To lead a lesbian or gay life is expensive because it usually requires independent housing, because meeting others is mediated by commercial services (bars, clubs, media), and because the subcultures are strongly sexualised, putting a premium on having the 'right' body and clothes. Gay life in present-day society thus suffers particularly sharply from the wider problems of commoditisation (section 6.5 and 6.9). Constraints on living out one's sexuality are therefore strongly material, worst for the poor and for women (Chauncey, 1994), and strongly variable between neighbourhoods. This form of social exclusion thus melds class and gender oppression in complex ways. Both oppression and freedom are constructed through relations between the spaces of the home, neighbourhood and city.

The elderly and people with disabilities

Deprivation of the elderly and people with disabilities is rooted in their marginalisation from the central relation of capitalist society, wage labour (Onyx and Benton, 1995; Oliver and Barnes, 1998). It is not merely that they receive little wage income but that their social reproduction is of no interest to capital. To the extent that capital's demands dominate state spending (section 5.2), state incomes and services for the elderly and support for their unpaid carers are weak. To the extent that dominant ideology in capitalism judges people's importance by their production of value, they are made invisible (Russell and Malhotra, 2001). Given weak state support, the living standards of the elderly are crucially dependent on wages over their working lives, and to some extent on that of their children, resulting in sharp differences in living standards among the elderly (section 5.8). Moreover, care of the elderly and the disabled by their family has been weakened by the impulsion on couples to have two wage incomes and to migrate for work (sections 6.2 and 6.3). Support and sociability are thus crucially dependent on the very variable strength of neigh-bourliness and local welfare services (section 5.4). Social relations at the *local and national* level thus have the effect of substantially confining to the space of the *home* the elderly poor and those with serious disabilities.

Children

Deprivation of children derives strongly from that of their parents; the child's space is above all the home. This dependency is socially constructed. Capitalist societies regard children as the property and responsibility of their parents; *society's* role in bringing up children is secondary (Donzelot, 1981). It is therefore regarded as normal that children should be poor because their parents are poor. Children are often socially isolated in their home, minimising interactions with other children and adults; children without adult supervision in public spaces are regarded as problematic, often disgraceful. Both neighbours and the state are reluctant to intervene to protect children from cruelty by their parents. The services provided to children by the state tend to be too weak adequately to compensate for the disadvantages of their isolation and their parents' class position. Children's development is, nevertheless, influenced by the physical and social environment of their neighbourhood and its public services; but for children living in *poor* neighbourhoods these produce further disadvantage (sections 5.3 and 6.6).

Teenagers and young adults

In a sense, the oppression of young people is the converse of that of the elderly and disabled: precisely the process of making them workers disadvantages them. This is the age where people are *graded* for their future role in waged and un-waged work, and *socialised* into that role. In a few years they have imposed on them the full weight of the division of labour and are socialised into the 'appropriate' gender, skill and cultural capital. Those whom the system assigns to the bottom are socialised into low expectations, poorly-resourced caring work, disciplinary low-skilled waged work, and frequent unemployment. This process is made the more frustrating by the large degree of inheritance of position from parents and neighbourhood. Young working-class men often seek to compensate for these humiliations by asserting their worth *as men*, through behaviour directed against their neighbours, the law and employers; in this contest the neighbourhood often loses, employers and the state usually win. Young women sometimes find in motherhood a role with *some* social valuation, but at the expense of their position in the labour market. For working-class women and men, youth is then a period of enormous tensions, which often gives rise to self-destructive behaviours.

Black and minority ethnic people, recent immigrants and refugees

Racist ideologies are rooted in the world division of labour created since European mercantilism: for people in the imperial countries, Third World peoples appear as 'naturally' poor and economically-backward (section 2.2). Since the nineteenth century, the rich countries have organised immigration to meet employers' demands for cheap workers for low skilled jobs; they have been willing to come because of under- or anti-development of their home countries. (The status of African Americans as former slaves has been worse, but unique among the rich countries.) Immigrants are then, at least for a period, at the bottom of the hierarchy of labour. In some cases this position has been due to their skills not matching employers' demands. But employers have also chosen to discriminate, and sometimes been pressured into doing so by white workers; they have been *enabled* to do so by BMEs' lack of options (Modood, 1997; Fletcher and Gaspasin, 2002). The state exacerbates these processes through immigration controls and obstacles to refugees, both fuelling popular racism and enabling the super-exploitation of illegal migrants (Hayter, 2004). Conversely, the most upwardly mobile immigrant groups were generally of higher class (better-off farmers, middle class) in their country of origin. Racist exclusion in employment thus involves capital dynamics and class relations *linking* the scales of the world, nation, and sometimes locality.

BMEs suffer from poor social reproduction by virtue of their lower incomes. This is intensified by racism in the general population which excludes them from better housing and sometimes social facilities, and which results in poorer treatment within parts of the public services (sections 5.3 and 6.6). This racism is based on perceptions of competition for local resources (section 7.2), thus imaginatively linking local spaces with the foreign. BMEs' materially inferior (but often socially superior) social reproduction then further disadvantages them in employment in vicious circles. As with sexism, racist exclusion arises from the *nexus* of production and social life, from both capital and competition among workers.

The relations of these different forms of oppression to class and poverty, then, articulate forms of power and use space in very *different* ways. But in all of them production relations play an important role. In some cases the problem is employers using the particular group for inferior jobs, in others it is the group's exclusion from employment; scales from the workplace to the globe are constitutive here. In all cases social reproduction also plays an important role: the private nature of the family and power within it, commodity consumption, and disadvantage in public services, mostly enacted at the scales of the home, neighbourhood and locality. We saw in section 3.5 that the *immediate* experiences of social oppression arise variously in production, social life and state welfare; we now see that the nexus of these spheres and the intersection of their spaces produce these deprivations.

7.4 NATIONAL REGIMES OF SOCIAL EXCLUSION

This part of the book has combined discussion of processes common to all developed countries with attention to the British case. But poverty and exclusion differ sharply between more-developed countries in their quantitative extent, forms, and social and spatial incidence (Mingione, 1996; Dixon and Macarov, 1998; Andersen and Jensen, 2002; Gordon and Townsend, 2000; Mayes *et al.*, 2001; Pinch, 1999). How would we seek to understand these differences? We do not have space here for even a sketchy analysis. But the discussion above suggests an *approach*, which we now specify briefly (for a fuller account see Gough, 2004a; Atkinson, 1999; Moulaert and Cavola, 2004).

We propose that inter-national differences in poverty are crucially dependent on relatively durable differences in economic, social, cultural and political practices of those national societies, which we shall refer to as 'national regimes' (Coates, 2000; Dore, 1997). National regimes are marked by their insertion in and use of space. We may picture them as provisional but they are often long-lasting settlements of the conflicts and tensions of class, gender and ethnicity. These national regimes have a number of aspects particularly relevant to poverty, which are connected, but not *reducible* to each other, and whose general forms we have examined in this part of the book. They include:

- *The social regulation of the economy and employment* (sections 1.3 and 1.4; Chapter 4). This comprises forms of capital, competition and co-operation between firms, industrial relations regimes, industrial strategies, and monetary and fiscal orientations (Aglietta, 1979; Bowles and Edwardes, 1993; Kotz *et al.*, 1994). One can contrast the liberal model of the USA and Britain, the Rhenish model of northern continental Europe, the Japanese corporatist model, and the southern European model. These crucially affect poverty through their long-term impacts on unemployment, security of employment, wage differentials, training and skill levels, career paths, and gender and age divisions of labour. They affect the politics of poverty through the social role and influence achieved by workers (de Brunhof, 1978; Therborn, 1986; Weiss and Hobson, 1995; Crouch and Streeck, 1997; Bamber and Lansbury, 1998).
- *Public services and income support* (Chapters 1, 2 and 5). The socialisation of reproduction through the welfare state, too, varies strongly between countries (Esping-Andersen, 1990; Mitchell, 1991). Its extent and quality has been higher in western Europe (including Britain) than the USA, Japan and southern Europe. The welfare state makes crucial differences to the quality of life of the poor, the caring work they have to do, the number of children per woman, the possibilities of independence from family, ability to access employment, and the pressure on the unemployed to take up a job.
- *Household forms, gender roles and sexualities* (section 6.2). Household forms and associated 'gender regimes' vary strongly between countries (Duncan, 1994; Duncan and Pfau-Effinger, 2000). Northern Europe, the USA and Australia have more non-family households and weaker extended-family ties than southern Europe and Japan. The freedom represented by diverse household forms is much harder to achieve if you are poor. Women's wage-work participation varies strongly by country due to gender relations

within the home but also whether the welfare state encourages or discourages it (e.g. Sweden versus Germany). Where wage participation is low, poorer women's independence is severely constrained; but a high participation rate can reflect not only choice (Scandinavia) but also pressures to compensate for low incomes (USA, Britain, southern Europe) (Perrons, 2000).

- *Systems of 'racial' exclusion and immigration* (sections 2.2, 4.3, 6.6 and 7.3). The most extreme social exclusion has been imposed on the indigenous peoples of the settler countries. Genocide, loss of land, forced migration, and suppression of traditional social organisation have led to very poor social reproduction (Pollard, 1988). In the USA, uniquely, the key role played by the descendents of the slaves in capitalism *from the first* has meant that this racialised divide has profoundly shaped the whole national society and culture; in consequence, the politics of 'race' and the politics of poverty have been really intermeshed *and* ideologically confused (Fainstein, 1993). The size of racialised communities created through immigration over the last century or so has varied sharply between countries – tiny in Japan and southern Europe, substantial in western Europe, massive in the settler countries – profoundly marking their national regimes. Larger immigration has given employers *possibilities* for control and division. External racist pressures have often led to the maintenance of excluding, exploitative and conservative relations *within* ethnic communities (Kakios and van der Velden, 1984). But racialised minorities have also played leading roles in struggles against poverty in general.

- *Inter-regional systems* (sections 2.4 and 4.3). The MDCs differ strongly in the proportion of the population in rural areas. Rural poverty takes different forms: high unemployment, low agricultural wages, or poor family-farms (D. Cox, 1998; Dirven *et al.*, 1998). Countries also differ markedly in the extent of uneven development between regions. This is largest in southern European countries and the USA (Dunford, 1994; Markusen, 1987). In the latter countries, especially, one may speak of qualitatively different regional *regimes*, for example between Catalonia and Andalucia or Lombardy and the Mezzogiorno. Large differences imply high levels of poverty and (often) disciplinary regimes in the backward regions, and allow employers in growth regions better to control their labour forces. They are an important source of economic-political division in the working class.

- *Urban systems* (sections 2.3, 2.5 and 2.6; Chapter 6). Housing systems and the social-spatial layout of urban regions, with their important impacts on poverty and exclusion, are strongly nationally specific. The English-speaking countries, with their liberal-individualist culture, have a particular dominance of owner occupation, disadvantaging the poor. Systems of residential separation differ markedly: in the USA and Britain the poor tend to be confined in inner city areas as the better off move out to low-density suburbs, whereas in other countries a high proportion of the poor are in outer estates. Countries also differ in the strength of state land-use and transport planning, roughly parallelling differences in the respective welfare states. Weaker planning results in worse accessibility to jobs and services and worse environments for the poor.

- *Ideologies of poverty* (section 7.2). The *dominant* attitudes to the poor differ strongly between countries, associated particularly with their production and public service regimes. The USA has the purest liberal ideology: the poor are losers who have got their deserts. In Japan, and to some extent southern Europe, with their acute senses of hierarchy, poverty may be shameful; but corporatist and clientalist patronage temper this. In contrast, in continental northwestern Europe poverty tends to be seen as a social rather than individual failing, though the social pact may not extend to women or migrants. British ideology, with its liberal production system but relatively strong welfare state and urban planning, falls somewhere between those of the USA and the Rhenish countries.

- **Collective action by the poor** (sections 7.2, 12.2). Finally, social underpinnings for collective action by the poor differ between national regimes. A societal ideology of poverty as collective rather than individual, and a strong welfare regime, encourages further demands on the state and (sometimes) employers (Offe, 1984). Organisational resources include strong trade unions (Europe, Australia) or rich extended family and community ties (southern Europe, Japan, racialised communities). The latter are also a basis for mutual support and maintenance of order (sections 6.3 and 6.10).

How are these national regimes created? Let us consider one example, Britain. The liberal production regime and individualistic culture of Britain have their roots in the early transition of agriculture to wage labour, and from the sixteenth century in powerful mercantilism. With the industrial revolution, Britain's primacy and its imperial markets allowed industry initially to flourish on the basis of fragmented firms, weak social organisation of innovation and skill formation, and 'distant' industrial relations. The Empire and Britain's political power beyond it encouraged financial capital to look overseas rather than form strong bonds with domestic industry (Fine and Harris, 1985; Ingham, 1984). The early production of an urban proletariat led to individualistic and market-oriented working-class culture, including in the spheres of gender and sexuality. But successful industrialism and large profits in both domestic and overseas markets also led to powerful trade unions and working-class pressure for improvements in living standards; in the twentieth century these led to the relatively strong welfare state and spatial planning. The Empire was also the basis for the racialisation of the division in labour.

This example suggests two points about the formation of national regimes. First, the relation between the national political-economy and the world order is of crucial importance. Second, regimes are developed through strategies of the different classes and conflicts between them. National regimes condition the creation of poverty. But reciprocally, these regimes are open to working-class struggles, in particular struggles against poverty and social exclusion.

CONCLUSION TO
PART II

We have argued that poverty and social exclusion are created by the nexus of economic, social and state processes, in which ideologies, culture and space are constitutive elements. Capitalist labour markets and their spatiality are central to poverty creation because they do not guarantee employment and because of the enormous differentiations in the quality of jobs which they produce; these impact also on those outside the labour market. The social relations of the economy also cause poverty-level state benefits both by restricting state spending and by linking benefit levels to low wages. But people's ability to survive on given money income is crucially dependent on the (spatial) forms of capitalist supply of goods and services, and on relations within households and neighbourhoods which organise distribution of resources, unpaid work, and reciprocal exchange. Survival also depends on state welfare services which, while they supplement money income and unpaid work, are limited in their benefits to the poor by their internalisation of class relations and social oppressions. Social reproduction both private and by the state in turn exacerbates differences in access to the labour market. Moreover, class disadvantage is compounded by, and compounds, social oppressions to produce highly differentiated forms of exclusion with their particular spatialities. All these processes are exacerbated by hostile attitudes to the poor and to poor places, attitudes which themselves have material-spatial roots. Labour markets, the private household form, their spatialities and associated ideologies continuously fragment the disadvantaged and set them against each other, making deprivation harder to cope with and to fight.

These processes of exclusion are characteristic of capitalist production, reproduction and state in the abstract. But they have evolved over time, through economic change and its impact on social life and the state, conducted through class and other social struggles. In nationally-differentiated ways, these caused changes over the twentieth century in the mechanisms of exclusion, culminating in rather sharp changes in the last thirty-odd years. The latter changes, in our view, have been generated not by technical or cultural change as such but by class struggle centred on an offensive by capital to restore both its authority and its rate of profit (see further, Chapter 9).

A corollary of this argument is that social exclusion does not arise from a failure of the excluded to conform to mainstream life, from biological, psychological or cultural inadequacy. Rather, *deprivation is systematically generated by the society as a whole and its spatialities*. It is true that some deprived individuals and communities have different social skills and social attitudes to the norm. But such differences are uneven and often absent; they are to a large extent functional survival mechanisms; and, most importantly, they are always *products* of society-wide relations of exclusion. Poverty and exclusion are, then, logical outcomes in capitalist societies. This is disguised by their variation by place and by social group; but this social-spatial unevenness is itself *an intrinsic part* of how the mechanisms of exclusion have their effect. It is also disguised by variation over time; but this does not show that capitalism can overcome poverty, but rather the temporal unevenness of capital accumulation, and ebb and flow in the class, gender, ethnic and other conflicts affecting poverty.

To say that poverty is necesssary in capitalist societies is not to say that it is primarily *designed* by a *collective* capital. It is true, as we have seen, that the state often deliberately sets out to lower incomes and to discipline the poor (Chapters 1, 2 and 5; see further, Chapter 9), but poverty is founded in the private decisions of firms and the private responsibilities of households. As a result, aspects of poverty may sometimes be *against* collective interests of capital, a problem to which the state may respond (sections 1.3, 1.5, 1.6; Chapter 2; see further, Chapters 8 and 10).

If exclusion is created by the society as a whole, then *the processes producing exclusion also oppress the majority of the population*. We have seen that these processes are not specific to the poor, though they have particularly severe consequences for them. Exploitation in work and the threat of unemployment are experienced by all workers; inequalities and conflicts within households and the difficulties of caring for dependents are widespread; problems with commoditisation of basic items of consumption, from housing and transport to food and the environment, impact on the majority; oppressions of gender, 'race', disability and age operate well beyond the poor; and the shortcomings of the welfare state and its disciplinary aspects are felt by all its users. This has important political implications. It implies that measures to alleviate poverty focused solely on the poor cannot succeed: effective policy has to address society-wide processes. If this conclusion induces pessimism, its corrolary should elicit optimism: the *majority* of the population – those mainly dependent on wage income and its derivatives – has the same interest as the poor in overcoming the processes which, in their condensed and extreme form, produce poverty. In the next part of the book we explore these political implications.

Part III

STRATEGIES AGAINST SOCIAL EXCLUSION

8

CONSENSUS STRATEGIES AND THEIR AMBIGUITIES

8.1 POLITICAL AGREEMENT AND DIFFERENCE

After the policy consensus over the post-war boom broke down, the dominant strategies for combating poverty and social exclusion changed dramatically. In this chapter, we outline the subsequent policy consensus for Britain; similar changes can be found in other countries, though conditioned by their particular historical and contemporary class relations (section 7.4).

The new consensus is explicitly hostile towards the key anti-poverty policies of the 1950s and 1960s. It rejects that

- national government should play the central role in combating poverty, for the most part through nationally-extensive policies;
- monetary and fiscal policy be used to promote full employment;
- national state policies should maintain output and employment in declining industries and that private investment should be steered towards weaker regions;
- protection for, and the rights of, employees should increase;
- universal welfare services should be extended quantitatively and qualitatively; and
- state transfer incomes should rise at least alongside average wages.

Although these policies might have been appropriate for an 'industrial' and 'Fordist' economy organised within national boundaries, the new consensus is that they are not suitable for a service-dominated, 'informational' and 'flexible' economy with strong global flows; that they involve too strong and direct a role for the state; that they are unresponsive to social and cultural diversity; and that they stifle enterprise and initiative. Thus, the starting point for the consensus is that the policies of the boom either failed or are inappropriate to the New Times (Hall and Jacques, 1991).

The new strategy seeks to relieve poverty indirectly by making the poor themselves responsible for improving their situation by acting more effectively in free markets. The strategy centres on integrating those of working age into wage labour. This is to be done not through job creation but by making the poor more employable. This potentially involves not merely training but promotion of an 'enterprising' character, a change in the culture of the poor, and reconstruction of social discipline, family and community. Poverty and social exclusion are seen as strongly locally specific, dependent on local labour markets and local social reproduction; policy application thus shifts from the national to the local scale. At the same time, the responsibility of the state changes: rather than intervening directly to ensure full employment and provide welfare services and income transfers, intervention shifts substantially towards quasi-state forms, partnerships and networks. This shift from government to 'governance' is preferred because it allows greater social and geographical flexibility in policy implementation, it removes a cause of 'welfare dependency', and it encourages or compels the poor to be active in improving their lot. Thus, rather than the national state overcoming (or compensating for) poverty, the poor and excluded are to be empowered to do better in rapidly changing, 'flexible' and 'globalised' markets. Whereas the post-war settlement promised greater (though not complete) *equality*, the new consensus promises *equality of opportunity* for the poor to compete in markets; greater substantial equality is not considered either possible or desirable (for a critique see Baker, 1987). New Labour became famously 'relaxed' about the amassing of wealth. In this chapter we examine the central elements of this new consensus. The assumptions of this consensus are that

- the unemployed are to be pushed into waged work;
- a central place is accorded to training;
- there will be spatial decentralisation of governance;
- there will be a concentration of intervention into 'the worst areas', particularly to build community capacity;
- support will be offered for the 'social economy' or 'Third Sector', neither state nor privately owned;
- poverty reduction has to be linked to environmental improvement; and
- organisation of these programmes will take place through partnerships and policy networks rather than by the state alone.

We shall argue that the shift to this new consensus has been driven by the neoliberal practices and ideologies which we discussed in Part II: removing impediments to 'free markets', allowing freer movement and decision-making of capital, increasing the disciplinary power of business and the state over labour and citizens, and cutting back and fragmenting the welfare state. Neoliberalism has attacked the post-war social democratic reliance on the state, and has emphasised the responsibility of individuals to provide for themselves and their families – the key assumptions of the new strategy. But, perhaps surprisingly, *the dominant approach has support from across most of the political spectrum*. This consensus makes it appear as apolitical, as common sense, reinforcing its legitimacy. In particular, these orientations are shared by two political strategies of the centre-left whose historical evolution we sketched in Part I.

- 'Conservative interventionism' is based on neoliberalism but seeks to enhance the economic, social and moral reproduction of the labour force, to improve the delivery of some public services, and to increase some forms of coordination of business.
- Associationalism aspires to diversify and deepen networks of actors in civil society, and hence introduce greater democracy into economic, social and political decision-making.

The period 1975 to around 1990 was one of neoliberal dominance, particularly at the national level. Nevertheless, during this period there were myriad *local* initiatives of a centre-left hue, and some left or associationalist ones (Boddy and Fudge, 1984; Eisenschitz and Gough, 1996). From the early 1990s anti-poverty policy shifted towards conservative interventionism, a shift consolidated by the election of Labour in 1997.

8.2 WELFARE TO WORK

The new consensus centres round the individualistic idea that, for people of working age, wage work is always superior to receiving income from the state or even from family. This is not merely because wage work can provide a higher income, but also because dependence on others for income is at best demeaning and at worst parasitic. The unemployed must therefore be pushed into work (Dean and Taylor-Gooby, 1992). But despite the importance of a job, policy does not seek to ensure that jobs are available. The 'globalisation' of the economy is said to severely limit the ability of nation states to ensure full employment, both because international financial markets constrain monetary and fiscal policy and because transnational ownership limits the ability of states to co-operate with business in saving or creating jobs. On the other hand, it is argued that investment is sensitive to the qualities of territorial workforces, and can therefore be *influenced* by improvements in the quality of such workforces (Giddens, 1998). Thus, the onus is decisively on the individuals to make themselves employable and thus attractive to business, rather than steering business towards the unemployed. Employability is to be enhanced first by people adjusting their expectations of wages and conditions of work to what is available, and second by improving their labour power.

Central to the delivery of this strategy has been the changes in the administration of state benefits and associated measures for the unemployed towards a system of 'welfare to work' or workfare (section 5.9). We have seen that in Britain since the 1980s benefits have been reduced relative to average incomes, entitlements to benefits restricted, and unemployed claimants pressed to take a job, participate in a training scheme or work in the voluntary sector. In Britain, as in the USA, work-readiness schemes have been substantially devolved to the local level, partly in response to the variability of labour markets between localities.

Workfare can, however, be carried out in different modes. In Britain and the USA it has for the most part taken a 'work first', disciplinary mode in which claimants are forced as soon as possible to take a job, however poor. This is consonant with pure neoliberal strategy. On the other hand, some local agencies have tried to pursue a conservative-interventionist approach, within which claimants are provided with substantial training and social support such as childcare or treatment for addictions, requiring higher state expenditure (Peck and Theodore, 2000). Associationalists seek to take this approach further by providing high quality training that builds on existing formal or informal skills. The centre-left also sees a major role for the Third Sector as a 'soft' jobs gateway (section 8.7 below). There are also significant differences in rhetoric. Pure neoliberalism threatens the unemployed with reduced or withdrawn benefits, treating them as 'economic people' motivated solely by income. New Labour, by contrast, has talked of the 'responsibilities *and rights*' of claimants and of them finding satisfaction and empowerment through work. However, 'rights' here do not include the right to a job, and may amount to no more than help with job applications and low-level training.

Conservative interventionism also differs from pure neoliberalism in seeking to make poor jobs somewhat more attractive to people living on benefits, and to somewhat improve their living standards when in work. The Blair government has thus introduced a minimum wage (though at a very low level and with important exclusions). It has also introduced tax credits for low-waged workers and for those using paid childcare to go to work. Pure neoliberalism sees these as a distortion of the labour market and argues that, if there is a poverty trap preventing people taking up jobs then the solution is to further reduce benefits. Blairism, in contrast, sees very low benefits and wages as leading to exclusion from social consumption norms and hence to dangerous political exclusion. But its solution is not to eliminate, but rather to subsidise, low pay, and not to make childcare free but to subsidise it, since the former options would be an excessive interference in the decisions of private businesses and families.

Across these variations, the consensus policy of workfare suffers from severe problems that we examined in Chapter 5. The most fundamental is that making people 'employable' does not necessarily lead to employment, particularly as the consensus has abandoned policies aimed at full employment. Moreover, the majority of jobs into which workfare schemes channel claimants are low quality, short-lived and do not improve skills. Workfare tends to exacerbate divisions between the unemployed: both (potential) employers and the schemes themselves grade participants by their existing skills, work experience and social group; and outcomes for participants vary strongly with local labour market conditions. Further, workfare schemes tend to worsen conditions in the lower end of the job market: by pushing more people into jobs, they put downward pressure on wages and conditions and increase the 'churning' of labour. Accordingly, workfare is not only an inadequate response to unemployment, but it also tends to worsen the employment conditions of the poor.

8.3 EDUCATION, TRAINING AND 'ENTERPRISE'

In the new consensus, education, training and the promotion of 'enterprise' are means to overcome poverty. If the poor of working age are to make themselves employable then education and training in the appropriate skills and attitudes to attract existing employers and new investors are crucial. This policy can enable the poor to compete equally in the labour market and hence to achieve according to their innate talents – a meritocracy of equal opportunity. Furthermore, there is broad agreement that training for working-class youth and for the long term unemployed should focus on inculcating work discipline and an aspiration to self-sufficiency through employment, values often called 'enterprise'.

Better education and training for the socially excluded tend to command broad political support because of their promise to benefit both workers – by enabling them to get (better) jobs and a sense of self-worth, and employers – by supplying them with a better quality workforce. This policy therefore tends to construct a cross-class alliance. Improving education and training to enhance employability is so much part of contemporary common sense as hardly ever to be questioned.

Thus, there is a broad consensus that primary and secondary education should prioritise the teaching of numeracy and literacy as competences distinct from any content, and prioritise technical, measurable skills in contrast to creativity and critique. The aspect of schooling as inculcation of abstract work-discipline, common to all capitalist societies in varying ways, has been accentuated in the present period (Bowles and Gintis, 1976). Links between local business and schools are seen as valuable in giving students a feel for the 'real world' of work, particularly for students who will not go to university. There is also consensus that state training programmes should be organised locally, an approach first implemented in Britain by the Conservatives with the creation of the local Training and Enterprise Councils in the late 1980s and continued (with somewhat larger 'local' areas) by the Labour government. As with work-fare implementation, the local scale is seen as allowing adjustment to very different labour market circumstances, given increased uneven development during the present period (section 4.4). The local scale also facilitates collab-oration between capital and labour over training since these negotiations, including the mediating role of training providers and the state, can be more direct and personal at this scale (see further, section 8.4 below).

There are, however, significant political differences on training, particularly for the unemployed and those without skills in demand. Neoliberal policy is to provide only the most basic training, equipping trainees only for unskilled jobs (Peck, 1993). This is partly because neoliberal theory focuses on cost rather than quality competition, so that policy is focused on swelling the labour force and thus holding down wages, rather than on substantial skills. Neoliberalism argues, besides, that higher level training, particularly that which is industry- or firm-specific, should be financed and organised by firms or workers themselves. Third Way strategy, on the other hand, holds that skills appropriate to the 'knowledge economy' are central to competitiveness, and therefore gives greater priority to 'serious training'. It is therefore in principle prepared to offer more than basic training to the unskilled. However, as we have already noted, workfare training under the present Labour government has in practice diverged only a little from the neoliberal model.

The belief in the efficacy of education and training in overcoming poverty suffers from the same problem as welfare-to-work: the absence of policies (or powers) to provide jobs. Unless improved training provision attracts extensive, job-creating investment to a territory, the *most* it can do is to enhance the job prospects of some at the expense of others. In practice, it seldom achieves even this redistribution, since the long term unemployed and those previously without any marketable skills tend to remain the last-chosen labour by employers despite training programmes. For the economy as a whole, extending education and training without altering production simply raises the paper qualifications needed to obtain a given job. Poverty consequent on unemployment and low pay can only be resolved by increasing both the supply and quality of jobs.

Moreover, the fetishisation of education as the route out of poverty has a corrosive effect on education as a whole by reducing it to job-training: understanding is reduced to information, and hence to competence in information technologies (Robins and Webster, 1988; Robertson, 2005) while artistic, imaginative and critical thinking, sport and exercise are sidelined. Similarly, the recent popularity of the 'learning' or 'informational' region privileges technical expertise at the expense of social understanding and the quality of social relations (Moulaert and Nussbaumer, 2005).

8.4 THE NEW LOCALISM: DECENTRALISATION AND SPATIAL FRAGMENTATION

A central element of the consensus is that poverty policies should be *locally varied* and implemented by *local actors*. In one sense, this spatial focus has a long pedigree. We saw in Chapter 2 that in the nineteenth and early twentieth century sanitary reform and physical renewal were focused on 'the worst areas', while state-led slum clearance and public housing programmes from the 1920s to the 1960s addressed poverty through neighbourhood housing change. But from the early twentieth century to the 1960s anti-poverty policy was increasingly focused on *national* level policies of redistribution and, in the post-war period, full employment, corresponding to deepening national administration of the economy and public services. The consensus developed since the 1970s has sought to reduce this national role. First, regulation of the economy by capital and the state and the organisation of public services have been substantially decentralised and fragmented *from* the national *to* the local or smaller scales, with supposed benefits for the poor. Second, a large part of national anti-poverty policy is *targeted* onto the poorest or most physically derelict small areas. We consider these two strands respectively in this section and the next. This 'new localism' has become the framework for most anti-poverty policy: while spatial scale has long been central to policy (Chapter 2), in the new localism it is explicit.

Local variability and targeting of policy rest on the argument that inter-local differences in economy and society have increased in the long term, especially among working-class areas. It is argued that the 'classic' industrial area with large manufacturing employers and nuclear families has been replaced by areas with varied forms of service employment, diverse gender divisions of labour, a variety of household forms and cultures, specific ethnic mixes, and greater differences in housing quality. This argument seriously underplays the previous variety in working-class localities (for example see McDowell and Massey, 1984). It is however true, as we saw in Part II, that since the 1970s economic stagnation and neoliberal policies have tended to increase social and economic differences between localities. The consensus argues that inter-local *differences* (of whatever origin) require locally differentiated *policies*. Failures of anti-poverty policy since the 1960s are seen as arising from bureaucratic nationally central-ised policies and a universal welfare state that exclude local people and agencies, prevent programmes being tailored to local circumstances, and hinder their mutual coordination at a local level.

The assumption of greater inter-local variability has led to the spatial fragmentation of **public services**, in forms we have already noted in Chapter 5. The National Health Service and school education are becoming increasingly locally differentiated and their units of delivery made more autonomous. In the neoliberal view, this forces down costs and thus opens the potential for service improvements; in the Third Way and associationalist view, decentralisation of control of public services enables greater grass-roots democratic input; in both views, the poor stand to benefit the most since they are the most reliant on public services and presently most disadvantaged within them. We have also seen that administration of unemployment and welfare-to-work has been substantially decentralised, again using the argument that local labour markets have become increasingly varied. The fragmentation of public services is seen by the consensus as fitting with their increasingly international regulation, allowing transnational investment in them.

Another key element of the consensus is that **the regulation of the economy** should be shifted from the national level both *upwards* to the international scale and *downwards* to the local, workplace and individual scales. Ever-greater internationalisation of production locations, ownership, investment, money and trade are accepted, with the corollary of the increased power of transnational industrial and financial firms and of international state institutions whose remit is to enhance free economic flows. In the consensus view this internationalisation will reduce costs and prices and stimulate investment worldwide, with particu-lar benefits to the poor. A substantial level of migration from the Third World and former communist countries to the rich countries is also accepted on the grounds that, despite high levels of unemployment in the latter countries, immigrants help to prevent labour shortages in certain skilled jobs and hold down wages and conditions in the lower end of the labour market; the latter ignores and tends to exacerbate the problems of secondary labour markets.

The consensus is that workplaces and localities should be substantially freed from the shackles of national regulation. Wages and conditions should be set by management or negotiated with unions at the workplace, firm or local level rather than nationally. The national state monopolies in utilities should be broken up, preferably through privatisation. Localities need to compete more consciously with each other. While these policies are most sharply articulated by neoliberalism, they are shared by conservative interventionism and associationalism. It is argued that this spatial variability is integral to economic 'flexibility' and hence increased investment, innovation and growth (section 4.3), which are seen as particularly beneficial for the poor, whether they live in prosperous or poor areas.

Within this vision of the decentralised economy there are, however, important political divergences. Associationalists argue that competitiveness depends not merely on holding down costs but also on enhancing productivity and innovativeness through networks of firms, agencies and (sometimes) labour. The latter may be substantially locally enclosed, as in industrial districts or clusters (Storper, 1997; Scott, 1998). Conservative-interventionist policy has implemented these ideas, though in rather weak forms. Moreover, the centre-left argues that local economic policy needs to do more than reduce local costs: it needs to address market failings in the supply of factors of production, which may, depending on the locality, include labour power, land and premises, technical knowledge generation and diffusion, and management expertise. Thus, whereas neoliberals interpret economic 'flexibility' to mean cost cutting, for the centre-left it includes pragmatic measures and networking to improve productive efficiency (Eisenschitz and Gough, 1996; 1997). For the poor, the centre-left strategy has a greater promise of creating reasonable quality, stable jobs, and of workplace and local organisation which gives some voice to labour (Cooke and Morgan, 1998). The centre-left does not, however, propose that the state should directly suppress the sector of poor jobs by economic or legal means, since this would jeopardise the vital flexibility and decentralisation of the economy.

The consensus around decentralisation raises a crucial question: how can decentralised governance bring greater socio-economic equality between rich and poor? We saw in Part II that the geographically uneven development of economy, social reproduction and state services is inherent in capitalism and has been deepened in the present period. This is not due to poor areas being autonomous from the rest, but on the contrary to their insertion into and links to the whole. Decentralised governance may be able to give greater formal power to poor areas; but if transfers of resources from the national level are reduced this formal control is useless. Moreover, decentralisation tends to exacerbate competition between localities and communities. Since the poor have the least resources *for* that competition, they are likely to be the losers. Moreover, the competition of localities weakens potential collaboration between different poor areas and between poor and middle-income areas. Thus decentralisation can *exacerbate* the socio-economic and political fragmentation and spatial unevenness that are at the root of social exclusion.

8.5 SMALL AREA REGENERATION: LOCAL TARGETING AND COMMUNITY MOBILISATION

A second strand of the new localism is the targeting of a large part of national anti-poverty policy onto the poorest or most physically derelict small areas. We saw in Chapter 2 the long pedigree of this policy. In the present period it can be traced in central government policy from the Urban Programme and Housing Action Areas of the late 1960s, through the Community Development Projects of the early 1970s, the funding for run-down built environments under the Inner Urban Areas Act 1978, the Conservatives' property-led initiatives in the 1980s, the more holistic urban programmes of the Major and Blair governments, to the various Action Zone programmes and the Neighbourhood Renewal strategy of the Blair governments. Since the 1970s there have also been many initiatives by *local* actors directed at the poorest localities and neighbourhoods but there is not a little irony in that most of the new localism has been initiated from above.

The focus on small areas is justified by the argument that poverty is concentrated in particular places. In the 1970s and 1980s these were the inner areas of large cities. Since the 1990s the focus has been on particular neighbourhoods, especially large council estates and areas of catastrophic employment decline such as pit villages. These concentrations of social exclusion are seen as a result of mutual reinforcement of economic, social, cultural and political disadvantages. We have seen that there are indeed vicious circles of disadvantage within neighbourhoods (sections 6.3, 6.6 and 6.11), though poverty is also reproduced in all localities (section 4.3). The assumption that poverty is localised does, however, serve a politically useful role. It means that both state transfer incomes and universal services can be cut, thus reducing 'waste', while poverty is still addressed through spatially-targeted spending. Poverty and social exclusion can be presented as a minor problem (if still important), theorised as a product of exceptionally inadequate individuals dumped together by the housing system, or of localised cultures of poverty, or of unusually poor physical environment, and thus as unconnected to mainstream society. Moreover, by targeting resources on small areas, there is a greater possibility of showing their effectiveness. Much (though not all) of the public unease about poverty is focused on its most *spectacular* manifestation in 'the worst areas', with their ugly environments, apparent lawlessness, deviance and social breakdown, which can be seen as making the nation incoherent and besmirching its image abroad (*vide* worries in France about the outer-city estates). Spatially targeted policy can provide reassurance that something is being done.

There is consensus, then, around special programmes for the poorest areas into which local governance of economy and public services (section 8.4) can be knitted. But in the last 30 years this area targeting has been implemented in very different ways, which partly correspond to differences in overall political strategy (Eisenschitz and Gough, 1993). From the late 1960s to the late 1980s local regeneration programmes were strongly focused on the built environment, including housing, industrial property, offices, overall urban townscape, and redeveloping derelict land; this proceeded through renovation or conversion of old buildings and new build. These programmes claimed to be primarily aimed at bringing jobs to the areas and improving residential space, and thus ultimately alleviating poverty; in this, they betrayed a fetishism of the built environment and, more widely, belief that one factor was the key to poverty alleviation. By the same token, however, they were able to 'show that something was being done' in the simplest way, by changing the look of the areas: old mills became artisan workshops; Mrs Thatcher undertook a public helicopter flight over London Docklands exalting in the numerous construction cranes; cities such as Newcastle, Manchester/Salford and Cardiff have been able to claim (and be widely believed) that they are qualitatively renewed on the basis of a half-mile of waterfront offices, arts facilities and upmarket flats.

These programmes – including even the Conservatives' Enterprise Zones (EZs) and Urban Development Corporations (UDCs) – had a Keynesian premise: that state intervention was necessary because of imperfections in the markets in land and property. The EZs, however, had a strong neoliberal element, at least at the level of political rhetoric: cutting taxes and weakening land use controls in small, derelict areas would show that low taxation and weak state regulation were the key to stimulating investment and 'enterprise'. Both the EZs and the UDCs were also neoliberal in their claims that commercial property development would benefit the local poor: the stimulus given by the programmes would establish higher levels of employment and an improved built environment which, through the *markets* in property and labour, would benefit all, including the poor (see further section 9.1).

By the early 1990s, however, both the focus on the built environment *and* neoliberal notions of local regeneration were widely discredited. Upgrading buildings and environments in inner cities and providing rent subsidies had failed substantially to address the job deficit. The firms occupying the new premises often had many weaknesses which better accommodation did not touch. More spectacularly, the EZs and UDCs failed to provide many jobs for local working-class residents. Both EZs and UDCs depended on enormous public subsidies for land reclamation and transport infrastructures. Much of the investment in them was in offices in which the majority of jobs were socially inaccessible to working-class residents. The UDCs stimulated the development of expensive flats and upmarket consumption, spaces that cashed in on the fashion among yuppies for central city locations; thus local working-class people were also excluded as residents and consumers (Colenutt and Tansley, 1990; Oatley, 1998). Neoliberal 'trickle down' benefits were small relative to the huge public cost.

More widely, by the early 1990s it was also evident that 15-odd years of neo-liberalism had created large numbers of areas with very serious levels of poverty and social exclusion; the most that could be said was that free markets had not *yet* improved them. Thatcherism tended to view this social destruction in the same way as it viewed economic destruction: as a painful but necessary adjustment to new conditions, as *creative* destruction. The social, cultural and political fabric of old working-class areas needed to be destroyed in order to enable labour in these areas to be moulded into new work routines. Meanwhile, there would be problems of disorder – petty crime and vandalism, riots of poor white and black men, large scale political protests as in major strikes and the poll tax rebellion; these would be dealt with by increased surveil-lance, expanded policing and the quasi-militarisation of the police. This pure neoliberal localism was therefore one of containment. But by the early 1990s a large part of elite opinion felt that this brand of localism was wasteful of resources and politically dangerous. There were high direct costs for the state in maintaining the unemployed and poor, in increased policing, and in increased demands on public services; sunk capital in housing and physical infrastructures was being wasted; the unemployed were socially and geo-graphically isolated from employers; and individual and collective disorder, and the levels of poverty they pointed to, were putting the legitimacy of neoliberalism in question.

These increasingly obvious problems led to important change in central government's urban programmes and to a strengthening of locally-generated regeneration initiatives. The Major government launched a new type of urban programme with City Challenge, which set the pattern for the Single Regeneration Budget and Labour's New Deal for Communities. During the 1970s and 1980s there had been numerous *locally* generated initiatives addressed to economic, social and political problems in poor areas, aiming to rebuild shattered communities, initiated by local authorities and other local actors. In the 1980s these had been severely constrained by national govern-ment; now they were cautiously encouraged. They received the blessing of the Labour government in its Neighbourhood Renewal Strategy, which sought to stimulate and multiply local regeneration networks through vehicles such as Local Strategic Partnerships.

The new approach aimed to be more **holistic** than the previous focus on the built environment. Local regeneration was to address the whole variety of problems facing the poor – employment, housing, transport, environment, education, crime and so on (section 3.4). Policy aspired not merely to address each of these but to address their interconnections, in particular by seeking better to coordinate branches of the local and national state through 'joined up government'. Overcoming poverty and social exclusion are seen as requiring advance on many fronts simultaneously. A component of the new strategy has been the neo-Keynesian approach to the local economy described in the last section. Both sticks *and* carrots have been used to strengthen social discipline, in particular parental guidance and control of youth. Neoliberal fragmentation of public services has continued, but has been given a twist: communities, rather than merely professionals and businesses, are to play a part in forming policy for these services and community self-help has been incorporated into an increasingly diverse spectrum of forms of delivery of social services (see section 8.7 below). As a result of this shift, the key measures – the spectacle – of 'successful regeneration' of poor areas have shifted: whereas neoliberal success is measured in business investment and construction cranes, the success of conservative interventionism is measured in less crime, less litter and vandalism, attendance at community meetings, participation in elections and in officially-compiled and publicised measures of progress in many aspects of life.

An enhanced role for '**community**' has been central to the conservative interventionist approach to local regeneration. In neoliberal strategy community is of little importance: as Mrs Thatcher famously said, 'There is no such thing as society, only individuals and their families'. Indeed, as we have noted, neoliberals have been happy to see the destruction of working-class communities and the solidarity that they sometimes offered. Neoliberalism uses the rhetoric of community in two ways. First, local identity is used to promote the economic competition of localities with each other. But this is a weak notion of community in that no non-market ties *between* community members are implied. Second, relations within communities are used either to strengthen social order, as in community policing and neighbourhood watch, or to mobilise soup-kitchen level self-help and voluntary action to address the worst manifestations of poverty. Conservative interventionist strategy, in contrast, aims for a stronger version of community. Relations within neighbourhoods are seen as important for strengthening reproduction within households, as supporting better education, training and (particularly) work ethic, and as the basis for strong and extensive forms of community business. Community representation is seen as an important input to local political structures in order to overcome the indifference or hostility of the poor to the local state (section 8.10 below). New Labour seeks not merely to *use* community ties but to *construct* them in order to counter anomie and social disintegration (Levitas, 1998): community 'capacity building' is a central thread. This positive vision of community is expressed in a particularly strong way by associationalism.

In conservative interventionism and associationalism, localism and community are thus more embedded in local society and more active than they are in neoliberalism. But they have their limits. Holism and joined-up government cannot go so far as to become the social-economic *planning* of post-war social democracy. Community organisation cannot become a mobilisation of neighbourhoods *against* the state or business, as in socialist strategy. The solidarity of localities and communities must remain subordinated to individualism, private property and markets.

Localism and community, then, have been prominent consensus themes, promising autonomy, regeneration and enterprise in the face of daunting national and international pressures. But they have been given quite different political colouring. This should not surprise us: the long history of the mobilisation of 'community' shows that this concept is always politically ambiguous and politically contested, lending itself to political projects from the extreme right to the extreme left (Crow and Allen, 1994; Mayo, 2000). These ambiguities are manifest in the contemporary consensus.

The concentration of intervention against poverty onto and within the poorest areas raises critical problems of *scale*. We saw in Part II that poverty is generated at different, interlocking spatial scales from the home to the globe; how then can intervention limited to a neighbourhood or locality address the problem? Small area policies have indeed often shown themselves to be on too small a scale to be effective:

- With regard to employment, in the EZs rents rose as landowners appropriated the tax breaks, and many new jobs were short-distance moves: space messes up projects of 'free markets' (Shutt, 1984; Massey, 1974). The UDCs achieved redevelopment of derelict land, but failed to recruit from surrounding working-class areas and thus to provide job benefits. The number of jobs in the EZ and UDC areas did, it is true, increase, since they were designated in locations potentially attractive to businesses. In contrast, the more 'holistic' urban programmes since the 1990s have largely failed to create more employment within their areas since these have been designated in poor residential areas unattractive to firms: job growth or decline has thus been determined elsewhere in the city. These programmes have therefore been forced to rely on improving poor residents' competitiveness in the *wider* labour market, with typically meagre results.
- With regard to housing, local regeneration programmes can improve the housing stock within neighbourhoods. But housing problems are substantially formed at sub-regional, regional and national levels (sections 6.5 and 6.6). Thus local programmes have no purchase on supply and prices in the (sub-) regional housing markets; nor can they address national underinvestment in social housing.
- With regard to education, neighbourhood programmes leave the social segregation and differential performance of primary and (largely) secondary schools across the locality and sub-region untouched.

Moreover, where small area policies have had greatest 'success', the poor have sometimes been excluded from this success and even displaced by it. The EZs and UDCs have directly cleared remaining industrial jobs and some working-class housing and leisure facilities through demolition, or indirectly displaced them through rises in land price. Some small area regeneration programmes have deliberately incorporated an element of gentrification, resulting in expulsion of poor residents (see section 10.4). To this extent, as with the clearing of the rookeries in the nineteenth century, the problem of poverty is simply spatially displaced.

8.6 ENTREPRENEURSHIP AND SMALL BUSINESS

One strand of the economic and cultural programme of the consensus is the promotion of entrepreneurship and support of small businesses of the poor. The 'rediscovery' of the small firm and its prominence in local economic strategies in the 1970s and 1980s (Storey *et al.*, 1987) has been translated into an important strand of consensus anti-poverty policy. The consensus around education and training includes not only enabling the poor better to compete for jobs but also equipping and encouraging them to form their own businesses. The promotion of an '**enterprise culture**' for the poor includes the capacity both to fight your corner in the job market and to employ yourself. Indeed, these are seen as involving similar aptitudes: the wage worker needs reflexively to deploy their initiative and to be prepared to change tasks, roles, skills and even industry, and thus to be flexible within a fast changing labour market; and hard work and formation of identity through work are seen as key to both employment and self-employment (cf. section 8.2).

Small business for the poor is promoted through training in business skills and through technical and financial support from small business agencies. In de-industrialised areas, such programmes are seen as a way of overcoming the shortage of employment by 'creating your own job', and also as a way of countering the lack of tradition of small-business formation deriving from the former domination of these areas by large firms. This policy is also seen as particularly helpful for groups which are disadvantaged in the employment market, particularly racialised minorities and men over 50 years of age.

Again, the different political currents see both the means and aims of small business start-ups in significantly different ways. For neoliberals, a high rate of start-ups is essential for driving inter-firm competition and thus driving efficient markets. Setting up a business is an assertion of independence and a reward for individual hard work, and is thus particularly valuable for the socially excluded in enabling them to overcome 'dependency culture'. A strong form of this neoliberal ethic is expressed in the promotion by the Black Muslim movement in the USA of black-owned small businesses as a central strategy for economic equality with whites, avoiding 'dependence' on white capital and state.

In contrast, the conservative interventionists and associationalists see business start-ups and development as needing *social support* as well as individual effort. 'Enterprise' is not solely or mainly an individual attribute but socially created. Neighbourhood or local networks of support are needed especially for the socially excluded, to provide them with the confidence, skills, material resources and social capital to make a successful business. The centre-left, moreover, argue that the *quality* as well as the number of small businesses is important. Simply increasing the stock of small businesses is not enough: the quality of processes, products, labour skills, marketing and firm organisation are crucial. These are to be enhanced through *local* networking, so that small firm development is a partially collective process; particularly for associationalists, cooperation and mutual support of small firms is crucial.

This strand of the consensus has many weaknesses (Weiss, 1988). The promotion of business start-ups without regard to their quality results in more firms being set up in minimally-innovative sectors, particularly consumer services, exacerbating the tendency to overcapacity in these sectors, and resulting in the failure of the start-ups or of existing (mostly local) businesses. Many of these businesses are no more than labour-only subcontractors, and thus in effect part of the secondary labour market (Rainnie, 1989). On the other hand, the poor are highly disadvantaged in developing innovative businesses due to lack of technical expertise, finance and contacts; most of the successful dotcom start-ups, for example, have been set up by people with university degrees, access to venture capital from family or friends and extensive business contacts. The centre-left policies of support are insufficient in the great majority of cases to overcome this disadvantage.

8.7 THE SOCIAL ECONOMY

A key element of the consensus is the building of the 'social economy','Third Sector' or 'community businesses'. These consist of enterprises and institutions supporting them, which are neither conventional small businesses, since their aim is not private profit making, nor state-owned or wholly state-financed services, nor non-commercial, in that they typically depend partly on sales income. The sector's very broad political appeal lies in its distance from both capitalist firms and the state, which as we shall see, permits many variations and ambiguities. It is seen as providing the best of both worlds – the individual choices offered by the market as well as the social solidarity afforded by welfare. The social economy promises simultaneously to play three roles in overcoming social exclusion: the provision of 'soft' employment, the supply of services useful for the poor, and the building of community ties and local democratic structures – the economic, the social and the political in one.

There is a great variety of economic and employment forms within the social economy (Mellor *et al.*, 1998). Ownership can rest with the community, with a charitable trust, with the local authority (in municipal trading enterprises) or with the workforce (in worker cooperatives). They may be for-profit or not-for-profit. The product may be sold to individual consumers or may be supplied under contract to a branch of the state or a quasi-state institution. The enterprise may receive a subsidy from the state, business or charitable trust in direct form or resources in kind (personnel, buildings, equipment). Jobs may be conventional and (semi-) permanent, as in worker cooperatives and in the numerous professional jobs in the larger social enterprises and in the sector's infra-structural organisations. But the sector also operates with people on (partially) state funded training schemes and with unpaid labour; in Local Exchange and Trading Schemes (LETS) participants are not paid in money but in scheme credits, enabling them to offer services to each other in a form of organised barter (and to avoid the payment of tax). The boundaries with the private sector, the state, and unpaid community work respectively are thus blurred.

The potential employment benefits are correspondingly diverse. Social enterprises are generally labour intensive, and thus tap the only plentiful resource of poor neighbourhoods. The social economy can provide conventional employment and train people in technical skills. But it can also provide experience of employment for those who have little or none, in an environment without the pressures of conventional business, and can enable jobs to be designed to fit with, for example, the needs of people with various disabilities. It can enable individuals to work in a strongly individual-autonomous fashion, as in LETS, or in a communal fashion with a sense of solidarity, participation and collective decision-making (Oatley, 1999).

Organisational forms are similarly varied. Some small-scale social enterprises are completely independent. Other enterprises are part of national 'voluntary sector' organisations or charities and may resemble franchises. Some, such as the Housing Associations and Housing Cooperatives, have a large input of state funding, are quasi-governmental, and are substantially regulated by the state. There are also infrastructural development organisations such as the local Community Development Trusts in the USA, which incubate and sometimes own social enterprises. There are now many 'social entrepreneurs' who act as (paid) consultants to the sector, finding ways through the maze of different legal, organisational and employment forms, the subsidies from different levels of the state and charitable trusts, and feasible business strategies; these social entrepreneurs may take the initiative in setting up a social enterprise, sometimes as a spin-off from existing ones. Networks are thus complex and multi-scaled; but the local and neighbourhood scales are, overall, the most important for formation and support of social enterprises. In Britain, the density of the social economy varies greatly between localities, depending on local community and political actors and cumulative growth paths, with some sub-regions such as west-central Scotland now having very substantial Third Sector networks.

We saw in Chapter 6 how the commodification of subsistence goods is a major cause of deprivation. The Third Sector can ameliorate this problem through various paths: by providing important services at a low price such as repairs to durables; by providing collectively-consumed goods or services such as environmental improvements; by relating producers directly to consumers, as in community gardens supplying fresh vegetables and fruit; or by dispensing with (conventional) money, as in LETS. The quality of the product may also be higher by virtue of the softer production relations and degree of worker control: thus cooperatively-run care homes may provide better care than privately owned, commercial ones. The social economy can thus promise to at least partially reintegrate production and consumption, and the economic and the social. On this basis, an enormous variety of goods and services can be provided: social housing of quality and price superior to the private rental market, basic banking and credit at non-usurious interest rates, affordable and trustworthy childcare and parents' centres, literacy and language classes, conveniently-located and cheap fruit and vegetables supply, or rural transport services. Services traditionally supplied predominantly by the state, such as education, health, community care, management of public spaces, and basic job readiness and help with job search, may be contracted out to social enterprises rather than private firms. All such community businesses may be specific to particular social groups, for example a particular ethnicity, either as workers, clients, or both.

Given the heterogeneity of relations to market, state and community, it is not surprising to find political tensions and variety within the social economy. One such tension is between small-scale initiatives rooted in local communities and nationally based, professionally-run 'voluntary' organisations and quangos. These differ in their potential for democratic accountability; the former organisations tend to be more innovative and risk-taking. Among locally based organisations, there are divisions between enterprises primarily accountable to their workforce and those controlled by residents' organisations. Another tension is between the pressures to operate in a fully commercial or even profit-making mode on the one hand, and commitment to social aims and the use of state subsidy on the other. A further dilemma arises when social enterprises take over functions previously run directly by the state, with employment consequently transferred out of public service agreements; this may be seen either as regressive privatisation or as a progressive alternative to operation by private firms.

Finally, there are tensions around how ambitious the sector should be, how far it can and should trespass into the territories of capital and the state. One model is of isolated enterprises addressing a local problem in a pragmatic way; another is of national federations addressed to a particular issue such as housing or childcare. A more aggressive strategy seeks to build the social economy at local, national and even international levels as a self-conscious movement with its own political values. In this vision, trading, financial transfers and flows of expertise link the individual enterprises into local and wider networks producing strong synergies, thus enabling each enterprise better to combine commercial viability with social commitments. The Mondragon cooperative movement in northern Spain is sometimes seen as a model of this approach.

These various dilemmas can be partially mapped onto the broad political strategies we have been considering. Neoliberalism, conservative interventionism and associationalism share a commitment to a particular class reading of the social economy. Its position between capital and state, between commercial and social aims, and between the logics of production and consumption is seen as lending the sector class neutrality, a sense of political 'balance', and thus ultimately as fostering desirable class cooperation. This has produced a

of sports participation, and getting working-class children into the countryside. The fear was that the working class had health so poor as to undermine both work and military capacity. In Britain in particular, concern with the physical environment has played a particularly prominent role. These old themes have linked up with the new environmental movement's concerns with pollution, resource depletion and unhealthy food. In a rhetorical and imprecise fashion, addressing these issues in poor neighbourhoods commands a consensus. Improving the environment is seen as an *apolitical* good since it benefits both business – through improving image and thus opening up investment opportunities – and the poor themselves. Since the early 1990s this consensus has settled around the notion of 'sustainable development' – the slogan of the Agenda 21 programme. This concept has had a complex relationship with anti-poverty programmes. On the one hand, funds to promote sustainable development have to some extent been targeted on poor areas as being the most needy. On the other, the centre-left has argued that 'sustainable development' has to address *social* sustainability. 'Sustainability' does not mean social justice or equality but rather the avoidance of open conflict. In both these modes, the *legitimacy* of the green agenda has been exploited to address symptoms and causes of poverty. The environmental element of the consensus has been further reinforced by its local focus: the problems being addressed are tractable precisely because they are narrowly spatially limited, locally specific, and do not imply any serious inroads into business interests or majority lifestyles. Addressing *local* problems can develop the participation of residents themselves (cf. section 5). Most obviously, improvements in the physical environment are easily visible and show that 'something is being done'.

Nevertheless, the consensus around the environment hides important differences in political approach. Neoliberalism can live with the kind of programmes described above providing their cost to the state is small and that they serve to improve the employability of the poor and stimulate local private investment. A spatial strategy of neoliberalism is to make problematic public spaces (quasi-) private, so that they are improved by private initiative: the creation of 'defensible spaces' in housing estates is an example. But neoliberalism has problems with collective control of space by sizable communities, fearing that this could become oppositional. The main environmental policy of neo-liberalism, the taxation of and market in environmental bads, is not part of the repertory of anti-poverty policy. Conservative interventionism, however, has seen the environments of the poor as a notable failure of neoliberalism. Poor environments are not only an all-too-visible sign of deprivation but are attributed strong causal power in the perpetuation of poverty. The moralistic strain in New Labour is concerned to educate the poor into a more responsible attitude to both their environment (for example, to stop littering) and their own bodies (to eat healthily). Highly localised programmes combining modest state funding with community enterprise fit well with this approach.

It is not hard to see the limitations of the types of policy described above. First, the policies do not touch the ways in which mainstream state services

Figure 8.1 Inauguration of building of a hotel, central Newcastle, 2002. A typical neoliberal partnership: the leader of the council, executives of the hotel and building companies, and a director of the public–private Centre for Life and former director of the local Urban Development Corporation (source: Newcastle Chronicle and Journal).

cause bad physical environments for the poor: the residualisation and semi-privatisation of social housing, selling off of school sports grounds, and cuts in local government spending on parks and street cleaning (sections 5.4 and 6.5). Second, the role of private firms and consumers in causing the particularly bad air and soil pollution of poor areas is not addressed; nor is the construction of ill health among the poor by normal consumer commodity consumption and normal wage work (sections 4.2 and 6.9). To address these issues would require policy to confront business interests, would demonstrate how poverty is linked to *mainstream* processes affecting the majority, and would take anti-poverty policy well beyond the spatial limits of the neighbourhood. This indeed is the agenda of the green left, socialists and socialist-feminists. Radical greens emphasise the need for major changes in lifestyles and hence consumption patterns, including through exemplary alternative living (Pepper, 1993), sometimes within an associationalist perspective. Socialists emphasise the need for fundamental change in both capitalist production and state services to improve built environments and protect nature. Socialist-feminists argue for a reorientation of economic priorities towards caring work, the body and its symbiosis with nature. These currents pursue this agenda through public collective resistance such as the anti-road-building protests. Conservative interventionist and, for the most part, associationalist policies towards the environments of the poor are limited by their refusal to take these confrontational and politicised paths.

8.9 PARTNERSHIP: GOVERNANCE, NOT GOVERNMENT

We have seen that the new consensus privileges the local and neighbourhood scales. The political-institutional form through which this is pursued is 'partnership'. Rather than local policy, including anti-poverty policy, being formulated and implemented by national and local government as it was in the long boom, it is to come from varied networks within each locality comprising parts of the state, state-sponsored agencies, business, voluntary organisations and, in contemporary versions of partnership, community groups; 'governance' is to replace government. Partnership is said to avoid an excessively powerful, authoritarian and insensitive state by involving varied social actors, thus enhancing local democracy. The more social actors are involved, the more a *local* consensus for addressing social exclusion can be created. Moreover, networks can coordinate policy across different aspects of economy and society and thus be more *holistic* than the state (cf. section 5). Through these mechanisms, they can generate a strong sense of involvement in and belonging to the locality, a political and ideological dimension to the new localism. Local partnerships are now an integral dimension of anti-poverty strategy.

Partnership has evolved historically under a number of partially contradictory pressures. We have seen in Chapter 5 how the local state in Britain was weakened in the 1970s and 1980s. This was part of a general neoliberal assault on the state, and specifically a response to the growth and strong politicisation of local government in the 1960s and 1970s. Local government taxation and spending were restrained, its regulatory powers weakened, and some of its assets and employment privatised. This was achieved in a large part through the actions of *national* government and a strengthening of its controls over local taxation and policy, since business did not have sufficient involvement or legitimacy at the local level to change local government practice. It was also facilitated by the internal centralisation of local government. As part of these changes an initial, distinctly *neoliberal* version of local partnership was developed. Local authorities were pushed to collaborate more directly and explicitly with business than in the past, when their pro-business policies had been mediated and inexplicit; local taxation, land use planning and education policies in particular were to be 'pro-business'. This stance was often institutionalised through local growth coalitions involving, most crucially, the local authority and representatives of business. National urban policy, the EZ and UDC programmes in particular, set up centrally-appointed agencies consisting in the main of business executives which took over planning, infrastructure and development powers from the local authorities. These forms of neoliberal partnership supplemented the contracting out of public services to the private sector.

This new form of local politics, however, proved to have problems. First, there were problems of popular legitimacy. Local government was seen as shackled by central government restrictions on the one hand and by the too-visible influence of business on the other. Policies of the new agencies such as UDCs and TECs were not open to outside scrutiny, let alone control (Oatley, 1998; Peck, 1993). Polices thus lacked local-democratic legitimacy. One symptom was ever-decreasing turnouts in local elections: why bother to vote for powerless bodies? Second, local governance had become fragmented through decentralisation, privatisation and a proliferation of qualgos (quasi-governmental local organisations). Policies in one field took no account of connections with others; notoriously, for example, the property-focused EZs and UDCs did not consider labour markets or low-income housing (Colenutt and Tansley, 1990).

Conservative interventionism has addressed these problems, not by re-empowering and democratising local government but by developing a new form of partnership. This has two key components. First, community organisations, particularly those in poor neighbourhoods, have a substantial, institutionalised role in partnerships. Second, policy fields, units of service delivery and communities have to compete with each other for funding. This form of partnership first emerged in the national City Challenge programme in the early 1990s, which, rather than designating favoured areas, obliged localities to compete for central funding, and which required local bidding bodies to involve community organisations. This has been a model for both nationally- and locally-initiated partnerships since then. It fits with conservative interventionism's wish to involve, and indeed build, community of the poor (section 8.5 above).

This strategy has nowhere been implemented in a pure form: the state remains enormously important in all developed countries, as do a myriad non-market arrangements between firms, workers and collective bodies. Neoliberalism is therefore best seen as a strategy for change in a particular *direction* – 'neoliberal-isation' as Peck and Tickell (2002) have termed it. Such change has been undertaken in all the developed countries, though to varying extents. 'Actually existing neoliberalism' is therefore a much more messy affair than the pure neoliberal blueprint (Brenner and Theodore, 2002). Here we explore the direction of change which neoliberalism has attempted; we examine some of the complexity of its outcomes in Britain in section 9.4 and in section 10.1.

Neoliberal theory does not ignore the needs of the poor, but rather argues that its strategy is the best for eliminating poverty. First, a more dynamic economy based on free markets will achieve a higher rate of growth, thus reducing unemployment and raising all incomes. Second and more specifically, the disadvantages of the poor in employment, housing and services are seen as created, or at least exacerbated, by political distortions of the respective markets; free markets will produce better provision of goods and services for people on low incomes, and will empower poor individuals to participate more fruitfully in the labour market. All these benefits are to be achieved not principally through specific policies *for the poor* but rather through the application of neoliberal principles to *entire fields* of economy, society and state. This stategy is sometimes referred to as 'trickle down', since the poor are in a sense indirect or joint beneficiaries. 'Trickle down'is often used pejoratively by critics, to suggest that neoliberalism is unconcerned about the poor. But neoliberals can legitimately reply that the alleviation of poverty is *central* to their strategy in that it is to be achieved by applying the *basics* of their strategy across the *whole* society; conversely, it is policies specifically for the poor which are superficial and tokenistic. Let us then examine in a little more detail the intended effect on the poor of neoliberalism in four broad fields.

Employment

On employment, neoliberals make arguments at different (spatial) scales, macro, meso and micro. The macro argument is that the removal (or minimisation) of non-market arrangements will overcome the worldwide stagnation which started in the 1970s. Weakening trade unions can lower excessive wages which damage profitability and remove a barrier to technical and organisational change. Lowering taxes increases net profitability and enhances firms' entrepreneurial instincts and workers' incentives to work hard. Reducing subsidies to and regulation of industry, and removing barriers to international flows of money and productive investment, allow production to shift from unprofitable and declining sectors to profitable and growing ones (section 4.5). Such measures can revive investment and growth, thus reducing unemployment and allowing faster income growth; the poor, being the most deprived of jobs and income, would benefit from these changes more than any other social group.

The meso argument is that disadvantaged groups have been excluded from employment opporunities by non-market arrangements which have been developed historically to protect relatively priviledged sections of workers and firms; the removal of these arrangements and the free play of markets is therefore *more egalitarian*. For example, training in Britain until the 1980s was organised through industry-specific bodies controlled by the employers, unions and the state in corporate fashion, with programmes focused on formal apprenticeships. This tended to restrict both the kinds of skills recognised and the social groups enabled to acquire them. Apprenticeships recruited overwhelmingly young white men, many of whose fathers were skilled workers in the industry. The abolition of this training system by the Thatcher government and its replacement with an ostensibly more market-driven system therefore promised to ease entry into skilled employment by previously marginalised groups. Similarly, neoliberals point to the many historical instances of manual unions using their influence over hiring to exclude women or black people.

Another meso-level argument concerns poor localities. Neoliberals argue that the explanation for weak local economies with high unemployment lies in *non-market* mechanisms which have deterred investment in them. In old industrial areas, for example, neoliberals argue that state ownership or subsidy has kept alive production which is unsustainable in the long term and which blocks conversion of the area to new sectors; that strong unions have inflated wages, blocked technical change, and supported provision of obsolete training; that state industries and local government have hoarded development land; and that local authorities have imposed excessive taxation on business and workers. Moreover, national wage bargaining keeps wages too high in weak localities; neoliberalism favours decentralisation of bargaining down to the enterprise, locality, workplace and ultimately individual level, so that wages more closely match profitability (Plougmann, 2001). Ending these non-market arrangements would allow the comparative advantage of weak areas to come into play, namely a large pool of free labour and wages below the national average which, in the absense of institutional constraints, encourage inward investment. This investment mops up the local unemployment, so that wages then rise. Neoliberalism, then, argues that it is necessary to allow wage and price differences between localities to increase precisely in order that their economic health should eventually become more equal.

Finally, neoliberals make the micro argument that a free labour market improves incentives for the poor to get jobs and thus improves their self-reliance and self-esteem (Green, 1998), an argument we return to shortly.

Housing

Neoliberals argue that problems of availability, quality and price of housing for the poor arise because of state regulation and ownership, an irony since the state's role was ostensibly intended to benefit mainly the poor. In the private

Figure 9.1 Neoliberal regeneration: the Metrocentre viewed from Scotswood, Tyneside. The Metrocentre, once the largest shopping centre in Britain, was built in an Enterprise Zone in the 1980s. Disinvestment from surrounding areas such as Scotswood has continued. Very few Scotswood residents have been able to get jobs in retailing (source: Jamie Gough).

countries (USA, Britain) than in others (France, Germany); but the reverse was the case in the 1980s, and the major country with the consistently lowest unemployment rate remains corporatist Japan. Moreover, even in the USA and Britain real unemployment rates are substantial and are very high in many localities and neighbourhoods. Neoliberals argue that the reason for this is that there still remain many social and political impediments to free markets. This is true: in labour markets, for instance, there remain health and safety legislation, rules on redundancies, collective bargaining, and substantial taxation. But this raises the question of why neoliberalism, despite its dominance over three decades, has been unable to remove such 'non-market distortions', a question we return to below.

A different interpretation of continuing high unemployment then presents itself: that neoliberalism produces medium-term effects which raise unemployment, albeit unevenly. First, cost cutting and intensification of work do not always stimulate investment. Though all production is to some extent cost-competitive, much competes on quality, uses skilled labour whose wages and conditions of reproduction cannot be forced down indefinitely, and may require large-scale, long turnover-time investments in fixed capital and R&D (section 4.3). 'Low road' cost cutting may be insufficient to stimulate such production

(Storper and Walker, 1989). Indeed, to the extent that neoliberalism promotes *reliance* on cost cutting, it tends to divert firms and the state from investment in high level skills and risky long-term investments.

Second, most low skill, low wage jobs in manufacturing and in large scale, routine office employment (back offices, call centres) are potentially mobile (section 4.3). Low skill levels weaken constraints of labour training and experience; and many of these jobs are weakly linked to other workplaces, institutions and infrastructures. While there may be investment in these sectors in lower wage countries and regions, these jobs tend to be unstable, with frequent relocations to other parts of the developed countries and to parts of the Third World.

Third, neoliberalism has in certain ways encouraged the growth of the effective labour force, thus increasing unemployment, other things being equal. As household incomes have been squeezed and had more demands made on them, there is pressure for more members of households to do waged work (section 6.2). In addition, the logic of neoliberalism is towards free labour flows across borders in order to create free labour markets. This logic has been countered by other political considerations. But immigration from the Third World has continued at a significant level in western Europe and a high level in the USA. The expansion of the labour force has been offset by an increase in those unable to work at the required intensity (section 3.3) and by the disincentives presented by the poverty trap (see below). But overall, the number of those seeking work has been increased by neoliberalism.

Thus although neoliberalism has been successful in expanding the number of low paid, low skill jobs, these have been insufficient to eliminate unemployment. At the local level few deindustrialised areas, despite their lower wage and other costs, have had sufficient or stable enough investment to overcome high unemployment. The fault in neoliberal theory lies in its faith in cost cutting to stimulate investment.

A number of other failures follow. Social groups which are competitively weak in the labour market (section 4.3) continue to be excluded from (stable) jobs, as employers are able to pick from the low-skilled unemployed. Welfare-to-work has limited effect in areas lacking jobs into which to force the unemployed (section 5.9). Thus labour shortages and wage inflation in growth regions coexist with high unemployment elsewhere.

Poor wages and conditions of employment

In neoliberal theory an expansion of low wage employment in response to economic stagnation should eventually lead to a rise in wages as unemployment falls and as intensive investment raises productivity. But, especially in the USA and Britain, the experience has been that low wage employment has continued to expand absolutely and relatively (section 4.4). This is partly because, as we have just seen, unemployment has not been eliminated. But it is also because

There are equity problems with such measures, and policies tend to come too late. More importantly, pricing of bads has not been applied with anything like sufficient energy to prevent them. As with regulation of service providers, this is because neoliberalism has delegitimated the state being tough with business and consumers.

The worsening of crime

The increase in relative and sometimes absolute poverty under neoliberalism has been part of the reason for the increase in blue-collar property crime (section 6.10). Neoliberal ideology may also have played a role. Crime is a strong exemplar of the individual *against* society. Indifference to the victims of theft, even if they are themselves poor, is deepened by neoliberal propertarianism. The explosion of corporate theft and white collar property crime allowed by deregulation legitimates theft. Thus the neoliberal exaltation of private property naturally creates its negation.

For all these reasons, the promise of neoliberalism to empower the poor as workers and consumers has not materialised. On the contrary, it has increased unemployment, worsened conditions of poor jobs, lowered state transfer incomes, caused deterioration in many essential services, and hence made domestic and caring work more onerous. These in turn tend to exacerbate oppression and conflicts *among* the poor, particularly along gender, ethnic and generational divides.

This increase in poverty has, over time, caused problems not only for the poor themselves but also for business and the state:

- The poor are costly because of state benefits paid to them, and sometimes the extra costs involved in meeting the higher demands which they make on public services. To the extent that poverty causes property crime, it increases the costs of policing and of stolen or vandalised property. As an opportunity cost, the unemployed waste the resources previously sunk in their reproduction.
- The potential benefit to business of the unemployed to undermine wages and conditions and provide a quickly mobilised workforce, does not function well. As noted above, many of the unemployed live in the wrong place to be accessed by employers; many lack basic skills and work disciplines.
- The poverty created by neoliberalism presents direct political problems. Crime and disorder perpetrated by the poor can be (or be made) a major concern of the general electorate. Riots originating in poverty have economic impacts mainly on the communities of the poor themselves. But it is impossible for politicians to ignore them, since they instil fear in the non-poor population and can make it seem that 'society is disintegrating'. They thereby create pressure to 'fix the problem' quickly, which is both

ideologically difficult and costly; they produce pressures for old-fashioned industrial intervention and for immigration controls which contradict neoliberal principles. Finally, the poor may organise forcefully against their poverty, though this has been muted in the recent past (see further Chapter 12).

These problems have produced pressure on the state from business directly to address poverty or its symptoms, as we explore in the next chapter.

9.3 REINTERPRETING THE PROMISE AND FAILURES OF NEOLIBERALISM

Why is it that neoliberalism has not achieved its promise to reduce poverty, but rather ended up exacerbating it? Our analysis of its immediate failures suggests some deeper contradictions in the neoliberal project and its theory.

First, it is not 'perfect markets' which create fast growth but rather a high rate of profit, actual and expected, and a willingness of capital to invest. These conditions often require what we have termed 'the socialisation of production', a set of non-market relations which support productive investment – in short, *imperfect* markets (section 4.3). Thus merely reducing the cost of labour and other production inputs in a territory does not necessarily lead to a surge in investment.

Second, high-growth capitalism is not a movement from one equilibrium to another within which supply and demand in all markets are balanced. Capital constantly seeks *higher* profits through change in sector and location and by reinventing the rules of territories' socialisations (Mandel, 1978). Territorial economies are thus constantly destabilised, with resulting unemployment and uneven wage levels.

Third, neoliberalism treats 'labour' as if it were an ordinary, inanimate commodity which is produced if the price is the free market one. This neglects the fact that (potential) workers are reproduced through complex and variable combinations of commodities purchased, domestic work, public services, and social life (Chapters 5 and 6). Moreover, workers have to offer some degree of cooperation with the employer for the work process to function adequately, requiring appropriate cultural formation. Wages, even supplemented by state transfer payments, may not adequately achieve these aspects of reproduction. Because reproduction is a complex social process, workers are not spatially mobile in the way that neoliberal theory assumes, so that the unemployed do not move automatically to where there is unmet demand for labour nor the poorly paid to higher paid areas.

Fourth, an aspect of all these points is that the spaces of production and reproduction are more lumpy, specific and frictional than the theory of free markets implies (Massey, 1974). Socialisations of production and reproduction in particular territories, and their complex articulation across scales, provide many barriers to the hypothesised workings of free markets. The socialisation of production within large workplaces and firms similarly produces lumpiness in service delivery at neighbourhood and local scales.

Fifth, neoliberalism runs up against the logic of the state, which it cannot dispense with as easily as it would like. Interventions that might strengthen the organisation of production and reproduction and better reconcile their respective geographies are eschewed. Weak regulation results in private firms providing poor services, for the poor especially, and firms and consumers producing damaging pollution. The neoliberal state cannot stand up to business, with the paradoxical result that the state becomes subordinate to *particular* businesses and lobbies. Because of political settlements of very long standing, neoliberalism cannot simply abolish state transfer payments – people cannot be allowed to starve or die of the cold on a large scale. Neoliberalism tries to cut this expense by making benefits more selective; but this increases the distortion of the labour market by exacerbating the poverty trap. The deeply-rooted logic of state spending and intervention thus conflicts with neoliberalism's attempts to scale these back.

Sixth, running through all these points are the cultural contradictions of neoliberalism. The attempt to reduce culture to market exchange and possessive individualism runs up against myriad social and economic practices which do not work in this way, from trust between employers and workforce to family and friendship ties to forms of caring and neighbourliness (Culpitt, 1999). Moreover, possessive individualism often destroys itself, as in neoliberalism's epidemic of crime and fraud. The wilful refusal of neoliberalism to think about these aspects of human relations then rebounds onto it, as the ruling ideology and the business class backing it are judged to have undermined a humane society and thus lose their legitimacy, exemplified by the decline of support for the Conservatives during the 1990s.

Finally, given the memory of the forms of socialisation dominant during the postwar boom, neoliberal austerity inevitably elicits popular pressure for the protection of the economic, social or cultural fabric of national or local spaces. Thus despite neoliberal precepts, national states have implemented strong immigration controls, and have indulged in expensive, inefficient but politically-reassuring local 'regeneration' initiatives (Oatley, 1998).

Underlying all these tensions are the contradictions between free markets and efficient socialisations, between freedom for firms and productive cooperation, and between mobility and spatial embeddedness (Harvey, 1982). The failures of neoliberalism, however, are not resolved by 'softening' it or making it more 'socially aware'. As we argued in Part II, neoliberalism has been a strategy for capital aimed at overcoming the crisis which appeared in the late 1960s and early 1970s. It aimed to do so by raising the average rate of profit and by depoliticising the economy, fundamental requirements for the reproduction of capital. And indeed, it has had considerable success. Since the early 1980s average profit rates have risen substantially in the major countries. The collective organisation of both workers and residents has been weakened, and propertarian-individualistic culture deepened (Geras, 1998). These are important victories for capital which will not be lightly jeopardised.

Yet the lacunae of neoliberalism have weakened not just the position of the poor but capital accumulation as a whole. At the end of the 1990s' boom the average rate of profit in industry and commerce had revived only to around its level at the beginning of the long wave of stagnation (Duménil and Levy, 2001a). Moreover, at the beginning of the twenty-first century the relation between industry and finance is far *worse* than at the beginning of this period, with accumulation both dependent on, and undermined by, credit, a relation further complicated by its inter-national geography (Duménil and Levy, 2001b; Gowan, 1999). It is these problems in the profit-investment circuit that are ultimately responsible for the problems discussed in the last section: continuing high levels of unemployment and cuts in wages and public spending.

Moreover, neoliberalism's successes in depoliticisation have started to falter. This is partly due to problems in the profit-investment circuit just referred to. Partly it expresses the cultural inadequacies of neoliberalism which have weakened its legitimacy, in particular through revulsion at its effects on the poor, an example being the majority opposition to the poll tax in the late 1980s. The evident subordination of the state to particular corporate interests (Monbiot, 2000) erodes its legitimacy. Furthermore, in continental Europe since the mid 1990s there has been strong resistance by unions and social movements to neoliberal austerity, in particular in reaction to the policies used to create and govern the Euro (Bonefeld, 2001; Taylor and Mathers, 2002).

One consequence of these contradictions is that, as we noted at the beginning of the chapter, neoliberalism has not been implemented in a pure form. This is not due to lack of political will but to fundamental features of modern society which neoliberalism denies or opposes. The extent to which countries have gone in this direction is consequently varied, depending particularly on their international geographies of capital investment and their longstanding class relations (section 7.4).

Neoliberalism has thus failed to restore an adequate rate of profit and has run into overt opposition. The failures of neoliberalism for the poor are thus congruent with, and integral to, its wider failures.

9.4 RACISM AND RIGHTWING POPULISM AS A STRATEGY AGAINST POVERTY

Rightwing politics in the present period has been centred on neoliberalism, but it has by no means been limited to it. Organic conservatism has remained important, as we see in the next chapter. Taking organic-conservative ideas to their extreme, right populism, centred on racism, has become increasingly influential in the last decade. This politics is seldom considered as a strategy against poverty since it has negligible support among academics and the 'political class'. But it has strong actual and latent support within the working class, and is therefore important to examine.

We discussed in section 7.2 how competition in employment and housing markets and for state resources can lead to racist ideas and practices among working-class white people, including seeing black people and immigrants as the cause of their poverty. Neighbourhood social life is a vital medium in propagating generalised racist attitudes and white subject positions and also specific stories about the 'privileges' given to local blacks or the wrongs they have committed. But also crucial are the discourses and policies of the major political parties and the mass media concerning economic and political immigrants (Beynon and Kushnik, 2002). Immigration policies which severely restrict entry by people without 'key skills' and by family members of people living in the country imply that migrants are exacerbating unemployment and stretching welfare resources. Exclusion of asylum seekers in violation of international treaties implies that they are a burden and a threat, an idea propagated by politicians (Hayter, 2004). Since 9/11, Moslems have been demonised by the popular press and by indiscriminate police raids.

The strength of these ideas has been the basis for far right populist parties. Since the 1980s these have grown substantially throughout western Europe (Betz, 2002; Flecker, 2002) and in Australia; they form of part of the variegated extreme right in the USA; the spectacular electoral success of the UK Independence Party in 2004 is a recent symptom. Being weakly linked to the corporations, these parties are not constrained by neoliberal strategy, and indeed their main policies are at odds with neoliberalism. Seeking to appeal to the white working class, they present 'obvious', direct and immediate solutions rather than the long term benefits of free markets promised by neoliberalism (section 9.1 above). Their principal focus is xenophobia and racism. Poverty is attributed to black people, always portrayed as 'immigrants', who have taken jobs and welfare resources away from their white neighbours. These resources are seen as having been created over time by the nation and thus as not belonging to 'immigrants'. Moreover, the rightwing parties often argue that black people, particularly recent migrants, have received *preferential* treatment in the allocation of social housing, income support and funding of community groups. Indeed, in localities with a substantial minority-ethnic community, it is argued that white people have been *socially excluded* by blacks, not only economically but culturally: the dress, food, patterns of sociability and religious practices of the black community are portrayed as threatening – 'you have to wear a turban to live round here now'.

The programme of the extreme right is to end all black immigration for whatever motive and to forcibly deport all black people 'to where they came from'. This is ostensibly to be achieved through the police apparatuses of the national state; but it also encourages home-grown violence at a neighbourhood scale. Local racist gangs are sometimes organised by fascist activists. Violence, whether by the national state or local citizens, is immanent in this politics.

While secondary to their xenophobia, the extreme right parties also put forward economic policies. These propose the protection of the national economy from globalisation, sometimes, as in the case of the Liga Nord in Italy, specifically from the malignant powers of the EU and the USA (Betz, 2002). The decline of manufacturing and increase in unemployment are attributed to free trade and abandonment of industrial support by the nation state. National enterprises and jobs for 'nationals' should be protected by tariffs and active industrial policy, potentially uniting capital, the petit bourgeoisie and workers as in classical fascist strategy (Guérin, 1973). Keynesian industrial policies are supported against neoliberal globalisation, and their inherent nationalism made explicit. While these proposals are unfashionable, they are given weight by the rhetoric of the mainstream parties and press concerning the need for 'national competitiveness', a notion which is in fact incompatible with neoliberal theory (Harris, 1995).

Scale, both material and imaginary, is thus central to this strategy. The *national* employment pool, welfare system and 'traditional culture' are to be protected from erosion by *international* migration, trade and capital flows on behalf of the 'indigenous' citizens, whether by tariffs, immigration authorities or white violence. The threats to this imagined heritage are often perceived particularly sharply at the scale of the *locality* (jobs, welfare) and the *neighbourhood* (housing, culture). Most of the activity of the populist right is at the local level, and is often very successful there. Indeed, the national elite is portrayed as having been coopted by *cosmopolitanism*, thus betraying both the nation and local communities (Flecker, 2002: 219–20).

Given the weakness of left alternatives to neoliberalism, it is not surprising to find that the unemployed and the poor are disproportionately represented amongst voters for the extreme right parties (Sivanandan, 2001). We should not underestimate the appeal of the latters' ideas well beyond their electoral base, despite their intellectual thinness and formal incompatibility with dominant neoliberalism. Any credible progressive response to social exclusion has to be able to undermine the social structures which generate these ideas.

to 'join up' departments, institutions and aspects of socio-economic life (section 8.9; Russell, 2001). The legitimacy of local government is to be strengthened not by increasing its powers but by transforming it into an 'enabling state', working through consensual and inclusive partnerships; partnership is to force 'modernisation' on local government (Carley, 2000: 44). The local state is to be democratic by virtue of this engagement rather than through the electoral process.

- All these measures, together with populist law and order measures and exclusion of immigrants, are intended to rebuild the *legitimacy* of the socio-economic order (section 9.2). CI wishes the self-seeking worker and consumer created by neoliberalism to take on additional roles as *citizen* and *neighbour*, 'with responsibilities as well as rights', engaged with his or her local community (Faulks, 1998; Cabinet Office, 2002). This takes up Putnam's (2001) arguments that social capital and sociability have declined in the English speaking world particularly, but that other regions show that they can be revived. Rhetorical and legal campaigns are launched against objectionable public behaviours such as drunkeness, yobbishness, vandalism, noise and incivility.

This attention to public services, limited socialisation of production, inclusion of the poor, social responsibility, community, good behaviour and the legitimacy of the state are all longstanding concerns of **organic conservatism**. At the centre of CI is an attempt to remedy the cultural failings of neoliberalism, the tendency of markets to cause social disintegration (section 9.3); organic conservatism has long pointed out that this is a problem *for capital*. Levitas (1986) has argued that organic conservatism was a component even of the Thatcher government. It is an enduring feature of centre-right politics in countries such as Germany, France and Japan. CI in Britain is in essence an inflection of neoliberalism with organic conservatism, or Christian Democracy as Marquand (1999) puts it.

That neoliberalism has failings for business and state does not mean that these failings can be easily addressed. Neoliberalism aimed above all to de-politicise the economy; the kind of interventions undertaken by CI could threaten to re-ignite political conflict, by unleashing demands for more spending on public services and infrastructures and more state support for industry – a return to the 1960s and 1970s (sections 4.5 and 5.6). But CI has had room to maneouvre because of the success of neoliberalism in imposing social discipline. In particular, trade unions have been greatly weakened numerically and in their self-confidence by a series of dramatic and decisive defeats in the 1980s; social movements have withered as organised forces; and community politics has abandoned the kind of aggressive struggles over housing and social facilities seen in the 1960s and 1970s. The *continuing* neoliberal framework of CI ensures that organised popular action remains at a low ebb: the sharper exposure of private and public enterprises to international economic flows imposes discipline; the culture of individualism and the ideology of 'globalisation' undermine faith in collective action; and it is directly blocked by anti-union laws and increasing legal restrictions. CI, then, has been able to cautiously increase some elements of public spending and regulation without eliciting strong demand for more. It has been able to end the 1980s exclusion of councils and community groups from local politics without either of these actors becoming demanding. The changes made by CI have thus – so far – avoided stimulating militancy; and on the contrary, they have served to diffuse discontent with neoliberalism. CI is thus a new, always tentative and changeable, class settlement (Strobel, 1996).

We have argued that neoliberalism proceeds through a particular articulation of space centred on factor mobility and cost competition between places (section 9.1). CI retains this thrust, but introduces elements of additional, different spatial logic. It seeks to *enhance interdependencies within territories*: citizenship at the national and, especially, local level; community participation; joined-up local governance; regional economic networking. CI priviledges the *local* scale, from region down to neighbourhood, over the national as the best one at which to effect a rebuilding of social reproduction (though increased public service spending and tax credits have been national policies). This is because a locality-by-locality approach *spatially fragments* pressures from the population and thus minimises the danger of renewed politicisation. Social needs are to be addressed, but according to the 'particular' needs of the locality and subject to its economic competitiveness. This heads off any revival of postwar ideas of universal rights (section 5.1) (Eisenschitz and Gough, 1996; Gough, 2002). The *local* scale of CI's tentative renewal of socialisation is intrinsic to its political feasibility.

10.2 FROM THE UNDERCLASS TO THE SOCIALLY EXCLUDED

A major element in the development of CI has been a conceptualisation of the nature of poverty and the poor: whereas neoliberalism came to conceive of the poor as an 'underclass', CI portrays them as 'socially excluded'.

The fundamental thrust of neoliberalism was to attempt to *integrate* those at the bottom of the labour market into the mainstream. The unemployed have helped to hold down conditions in low skilled jobs and enable the expansion of this sector (section 4.2). Workfare is intended to facilitate these processes as well as to reduce state spending. Moreover, the poor were used to discipline the entire workforce: high levels of unemployment and deterioration of conditions in the lower end of the labour market warn those in better jobs: unless you accept employers' discipline, this is where you may end up. These mechanisms are in line with neoliberal theory, in particular in seeing everyone, including the poor, as 'economically rational' (section 9.1).

Yet this project comes up against neoliberalism's manifest failure to solve the problem of poverty. Unemployment has turned out not to be a therapeutic episode leading to economic revival but long term, albeit with spatial and temporal unevenness. It has been difficult to force the unemployed to take poor jobs, given the politically-necessary level of state benefits, the poverty of the jobs on offer, and the lack of material support such as childcare and transport costs. This political embarrassment is then covered by labelling the unemployed as work-shy and feckless, justifying increasingly disciplinarian workfare measures (Handler and Hasenfeld, 1991). Thus the poor are no longer portrayed as 'economically rational' but as *culturally* different from the rest of the population, that is, an 'underclass' (sections 1.5 and 2.6). Moreover, the neoliberal reliance on markets fails to *socially and ideologically* integrate the poor, manifested in crime, drug abuse and riots (section 9.2). Again, the theory of the underclass can transfer the blame for these problems onto the culture of the poor; and this then legitimates stepped-up policing of poor areas, 'wars' against drugs, curfews on youth, demonisation of lone parents, and punishment of 'bad behaviour' by welfare agencies. Moreover, we have seen that neoliberalism has seen its legitimacy for the *majority* of the population come under threat (section 9.3). A discourse which blames wide socio-economic failure on laziness and social indiscipline *in society as a whole* provides an excuse for failure, as in the moral panics around football hooligans and teenage pregnancies. This discourse of wide social indiscipline chimes with that of an underclass (though it paradoxically implies a *lesser* cultural distinction between the poor and the majority).

The theory of the underclass is often seen as a natural part of neoliberal ideas (e.g. Jones and Novak, 1999). But it is a departure from neoliberalism's 'economically rational people'; its focus on culture and social discipline is borrowed from organic conservatism. Neoliberalism has however used this theory to justify *coercive* measures against the poor by the state.

However, the plausibility of the notion of an underclass is undermined when, as in the last three decades, unemployment and poverty increase rapidly and hit 'normal' sections of the working class. Moreover, the disciplinary approach to the underclass has its contradictions. It can undermine the neoliberal aim of integration of the unemployed into the labour market: labelling the poor as 'dangerous' scares off employers, and coercion removes the self-esteem needed to find a job (Dean and Taylor-Gooby, 1992). Moreover, this authoritarianism clashes with a dominant ideology of late capitalism: the autonomous consumer pursuing his or her lifestyle choices without interference by the state or employers. Whereas the authoritarianism of the nineteenth-century ruling class was founded on traditional notions of class deference and self-discipline, a century of capitalist development has eroded these ideologies.

CI's conceptualisation of the poor springs from the manifold failures of neoliberalism: the deepening of poverty, the alienation and apparent political threat of the poor, their costs (sections 9.2 and 9.3), and the contradictions of purely disciplinary policy. CI has retained the theory of the underclass, whose cultural-political focus fits closely with CI's organic conservatism (Levitas, 1998). The concentration of New Labour's poverty strategy on 'the worst areas' (section 8.5) fits exactly with this theory: a *spatial concentration* of poverty, rather than poverty itself, is the real problem, since it has the critical mass to generate a distinct, disfunctional culture and a threat to order. But whereas neoliberalism deals with 'the culture of poverty' by repression, CI seeks to *reform* it through a mixture of economic, social, cultural and political interventions. The poor are not simply to be forced to work but rather helped into it. The supposedly poor upbringing of children is to be addressed not only through discipline by the police and schools but rather through putting extra resources into childcare and education. Community organisation in poor areas is to be encouraged in order to overcome their supposed anomie and build social capital (Forrest and Kearns, 1999; Cabinet Office, 2002). The supposed weakness of family life among the poor is to be addressed not merely through penalising 'improper' families through the benefit system but through giving tax credits to parents. The connections between the different aspects of deprivation – employment, housing, environment, schooling, health, crime – which *reproduce* an underclass are to be directly addressed. The underclass is therefore to be tackled not only through coercion but also through support designed to strengthen community, family, employability and citizenship.

Enter the concept of 'social exclusion'. If the underclass can be reformed, then the term itself is a barrier to that reform. Moreover, CI understands the underclass not as a series of unsatisfactory individuals but as *socially* produced. The culture of the poor is certainly a factor in their continuing poverty, but is produced by material practices which are susceptible to change; *inclusion* is therefore possible through appropriate *social* policy. The problem is also *social* in that it involves the interaction of different aspects of life – economy, family, community and so on, especially within neighbourhoods. This approach is inspired by the structuralist thinking of organic conservatism against the methodological individualism of neoliberalism. Moreover, the redefinition of the problem as one of social exclusion rather than simply of poverty shifts from the purely 'economic' concerns of neoliberal thought to a wider set of social relations: the problem may be not merely lack of money but patterns of life. The latter are seen as a major cause of poverty; but the moralistic strand of CI also sees them as in need of change *in their own right* (Strobel, 1996). CI's slogan of 'social inclusion' thus suggests that neoliberal coercion of the poor needs complementing by social engineering to change the patterns of their lives.

10.3 CONSERVATIVE INTERVENTIONISM VERSUS SOCIAL EXCLUSION

The aspiration of CI, then, is to reform the culture of the poor particularly in order to integrate them into the labour market. The aim is to create a meritocracy within which each has an equal *opportunity* to fulfill their potential freed from social constraint. It is claimed, in Gordon Brown's words, that there will then be 'fairness of outcome'; *equality* of outcome is regarded as neither feasible nor desirable. CI claims that part of its programme against social exclusion is through improvements in public services for the whole population

Worpole and Greenhalgh, 1999). Indeed, for old industrial cities in particular this strategy involves a comprehensive change in image: the end of proletarian industrial culture, the reinvention of the city as 'informational', high-cultural, a centre of 'café society'. If some waterfront can be found to convert from industrial and port uses, so much the better.

The strategy is presented as one for the whole city but also as a way of overcoming social exclusion. The economy is to be feminised through growth in service work. The new consumption spaces are to be multicultural. Zones of the centre are to be more cosmopolitan with a minority-ethnic stamp while young, childless professionals are encouraged to revive previously run-down parts of the city centre. The city centre should encourage mixing and thus greater tolerance, and 'celebration of diversity' (Allen and Cars, 2001). Thus the new production and consumption spaces are to be more sophisticated and more *inclusive* of previously excluded minorities, replacing the social divisions of the old proletarian culture.

The strategy of beneficial cultural mixing also extends to *class* mixing. The partial gentrification of some working-class districts is seen as not only helping to change the city's image but also as helping to improve the lives of the poor in those or in adjacent neighbourhoods (Carley *et al.*, 2000: 29). The middle class will improve both private and public services; their attitudes to work and education will rub off on their neighbours; their presence will overcome the stigmatisation of neighbourhoods. As we saw in Chapter 2, this strategy has a 200-year history. Whereas for neoliberals gentrification is an opportunity for the construction industry and the middle-class consumer, for CI it lessens class tensions.

Strategy to develop the post-industrial city has changed significantly over time. For Thatcherism, the decline of manufacturing and the rise of services in cities was driven by markets, while cultural diversification was powered by consumer choice. The new city was one of opportunity for developers and consumers. Government only had to ensure that change was not inhibited by state support for the old regime and that land was fully marketised. But, for CI, active governance is often needed to catalyse the transition. Old industrial regions and cities need to coax informational services in; place marketing is needed to convince investors, professional workers and tourists that the city has *already* changed (Kearns and Philo, 1993); and land use, transport and housing planning are needed to help change the city centre. Actual or potential conflicts between different groups within the city have to be managed. Longstanding businesses faced with eviction by rising rents or compulsory puchase have to be placated; conflicts between new city-centre residents and commerce have to be mediated; gay villages and ethnic enclaves have to be legitimated and protected from attack by working-class gangs; and gentrification has to be sold to the poor (Peck and Tickell, 1995; Taylor *et al.*, 1996).

As a strategy for social inclusion, the post-industrial city has severe problems (for the case of Glasgow see MacLeod, 2002). First, it is only possible to implement in the largest city-regions which have some chance of generating the minimum efficient agglomeration of high level services (Bryson *et al.*, 2004). Recently it has been claimed that the city can be the 'motor of growth' for the (sub-) region; but this simply points to the lack of credible policies for the revival of old industrial towns and villages in depressed regions. Second, the job opportunities which the new model affords to the working class have poor wages and security: routine office work, consumer services, lower-end public services, and domestic service (section 4.4; Fainstein *et al.*, 1992; Hamnett, 2002). Access to the glamorous creative jobs and self-employment is heavily biased against the socially excluded (Tams, 2002)

Third, the glamour jobs, new consumer services and gentrified housing are in central areas or priviledged suburban enclaves, so that the regeneration of the built environment is partial and uneven (section 6.6). Nevertheless, the gentrification of previously poor neighbourhoods tends to raise housing prices throughout the city-region, and evicts working-class residents either through displacement from existing properties (Smith, 1996) or through demolition to create sufficiently large, 'defensible' sites to attract private developers (Peel, 1993; Byrne, 2000).

Fourth, the new privately-run leisure spaces are relatively expensive and exclude the poor (Chatterton and Hollands, 2003). Most of the publicly-funded high culture is detatched from working-class life (Byrne and Wharton, 2004). Indeed, the aim of the strategy is to erase the image, if not the reality, of the city's traditional proletarian culture (Strangleman and Roberts, 1999). Within oppressed groups the benefits of the new consumption spaces are least for their poorer members. For example, the residential parts of gay villages, and to a lesser extent the bars and clubs, are dominated by middle- and upper-income men (Knopp, 1990); homophobia in other residential areas is not touched. Indeed, as gentrification increases housing costs it becomes harder for low income lesbians and gay men to escape their family households. Thus both employment and consumption in the new model city are strongly polarised by income. Overall, then, CI's post-industrial city strategy has weak benefits and a number of disbenefits for the excluded.

10.5 FAILURES OF SOCIAL INCLUSION

CI has promised to ameliorate some of the problems for the poor thrown up by neoliberalism. But the example of the post-industrial city suggests that its conservative-integrationist aims are undermined by neoliberal markets; we shall see that they also run up against the neoliberal reforms of the state. We can examine these tensions in a number of spheres.

Unemployment, poor jobs and low incomes

While CI sees employment as the key to overcoming social exclusion, it has no mechanism other than the further freeing of markets to ensure that **jobs** are available. To the extent that workfare forces the unemployed into jobs, this merely results in churning. Under New Labour governments unemployment has fallen rapidly. But this has been a result of an extended growth period of the business cycle involving a chronic balance of trade deficit and unprecentedly-low savings ratio; there is no evidence that a low-unemployment *regime* has been constructed. Moreover, over 2 million people are on invalidity benefit, and many localities and neighbourhoods retain high official unemployment. This problem radically undermines the project of CI: it not only prevents incomes from rising but undermines the social responsibility inculcated by work which CI regards as so important.

For the unemployed who obtain a job, the **quality of the job** is generally poor. The current government uses the notion of a 'knowledge' or 'information economy' to suggest that jobs are becoming more skilled and allow more autonomy to the worker. This is misleading. Even *skilled* informational jobs are in many cases being subjected to detailed division of labour, authoritarian supervision, intensification and casualisation (Smith, 1994; Morris-Suzuki, 1997). Despite the increasing importance of knowledge in production and of informational services and goods (Lash and Urry, 1994), a high proportion of all jobs, including most consumer services and much manufacturing and business services, remain low skilled, repetitive, low paid and vulnerable to casualisation; many of these jobs are also potentially mobile. We saw in Chapter 4 that the growth in employment in Britain since the mid-1980s has been largely in poor jobs. Indeed, this is a product of neoliberal strategy: the disempowering of labour allows job growth without excessive wage inflation. CI's policies for 'high road' production (section 10.1) are too weak to affect this trend substantially. Furthermore, disciplinary employment relations across the whole economy contradict the strategy of developing a *commitment* to work among the poor.

The **social economy** has the potential to ameliorate this problem by foster-ing self-managed enterprises with a high degree of responsibility for the workers and development of their technical and generic skills (even if the wages are low). But the mode in which CI has supported the social economy, while an improvement on pure neoliberalism, largely negates this potential. CI does not support the extension of the social economy to the non-poor, since this would move it into the territory of private business. Community enterprises are still treated as one-offs, expected to be largely autonomous, without the construction of larger networks which could strengthen them both economi-cally and, crucially, politically and ideologically. Ownership is not lodged with local communities. Social enterprises are therefore subject to market forces; even those receiving local government contracts are under pressure to cut costs. Moreover, CI sees the social economy primarily as a *stepping stone* for individuals to improve themselves and escape poverty. The participants are therefore not expected to stay in the sector nor in the neighbourhood in the medium term. This undermines the development of a social economy rooted in the community. This narrow agenda limits the potential of the Third Sector both quantitatively (number of viable jobs) and qualitatively (progressive relations of employment).

CI has introduced **policies against low wages**, but they are weak. The Minimum Wage has been set at a very low level (well below the EU's 'decency threshold'); it is weakly policed – in common with nearly all other regulation of firms – so that much casualised work paying abysmal wages continues. The level of the working families tax credit still leaves very high effective marginal rates of tax for those moving off benefits, especially for those in high rent areas receiving housing benefit (Millar, 2003b). These credits have been insufficient to reverse the regressive tax changes of the Conservatives: between 1979 and 2003 the proportion of income paid in tax of the poorest fifth of taxpayers has risen from 31 per cent to 42 per cent, while the proportion paid by the highest fifth has declined from 37 per cent to 34 per cent (Clark, 2004). These problems reflect CI's unwillingness to take on business and to use the tax system for redistribution.

The low quality of jobs on offer to the poor contradicts CI's cultural agenda. CI seeks a revival of the classical Labourist 'respectable' working class – social democratic in economics, conservative in social matters, believers in family, order and authority. But this class identity depended on skilled, stable, reasonably-paid male jobs (Thompson, 1968). The consignment of the great majority of the working class to poor jobs radically undermines the CI project. As Sennett argues (1998), the dominant contemporary models of work militate against a sense of social connectedness and responsibility.

The Labour government has also failed to improve **the distribution of income**. The slight reduction in the numbers of the poor from 13.9 to 12.4 million between 1997 and 2002, largely reflected falling unemployment. But the distribution of income among the poor widened (the income of the poorest 1 per cent actually fell *absolutely*), and incomes of the well-off continued to race ahead. Between the fiscal years 1996/7 and 2003/4 post-tax income inequality actually increased slightly (Brewer *et al.*, 2005). The continued widening in pre-tax income inequality was inadequately compensated by the government's redistributive measures, particularly due to the refusal to increase benefits and pensions for non-workers (sections 4.4 and 5.7) and to make the tax system more progressive.

It is true that CI professes to be unconcerned with the distribution of income, only with maximising equality of opportunity. But an enormous, and increasing, inequality of income inevitably undermines its strategy for the poor. The well-off can buy an increasing range of commodities, including superior education and healthcare, which are inaccessible to the poor and thereby exclude them *from the social norm*; and it is *inequality* of income, not merely absolute income, which prevents upward mobility (Clark, 2004). Thus even equality of opportunity is thwarted.

Poor and class-divided public services

CI's aspiration to improve public services for the poor, through 'mainstreaming' and through small-area programmes, runs up against the neoliberal reforms of these services which CI has continued and deepened (section 10.1). These exacerbate differences in quality of units of delivery, usually to the detriment of poor clients (sections 5.4). This applies also to services devolved to community enterprises: if the funding is insufficient, and if clients are limited to the poor, their quality will be low despite 'community control'. Marketed public services such as public transport and social housing have been further deregulated, resulting in poorer quality and more expensive services for the poor.

The continued widening of income differences under CI exacerbates these problems. The self-segregation of the better off residentially through their buying power in (increasingly differentiated) housing markets, and hence class segregation in schools, continues to deepen. Some public services are abandoned by people of middle income (buses everywhere except London) or by the middle class (education in inner London), weakening the political pressure to maintain them (Burchardt et al., 1999). Public services themselves, and their increasingly poor clients, both tend to become stigmatised.

Community, difference and power

CI's neighbourhood-community focus often runs up against the weakness of community organisation in many poor areas (section 6.3). Local partnerships then have difficulty ticking the boxes of 'community involvement' to obtain further funding. The real problem here is not obtaining the views of the poor: the needs and aspirations of different areas are not as varied as postmodern thought supposes, and easily obtained data on the area can indicate well enough what is needed. The problem is rather how the poor can exert effective political pressure.

Where there *is* existing community organisation, CI underestimates, and thereby paradoxically exacerbates, their internal conflicts, whether of gender, 'race', age, housing tenure, neighbourhood or class (McCulloch, 2004). CI's community regeneration approaches these differences in two modes. The first is to ignore them and assume a unified community interest. This allows stronger groups to sabotage initiatives which they dislike. In Scotswood in Newcastle, for example, attempts at improvement in policing, child care and health were carried entirely by women; they were opposed by locally-powerful men in criminal gangs who feared *any* involvement of *any* part of the state in the neighbourhood (compare the ideology of the US militias). The regeneration partnerships did not acknowledge this antagonism, and thus allowed many of the initiatives to be sabotaged (Campbell, 1993).

The second mode of dealing with difference is to allocate an identifiable tranche of resources to each interest group and claim that the programme is therefore 'balanced'. This reinvents the long tradition of local pork-barrelling politics and clientalism, dressed up with a postmodern 'celebration of difference'. It tends not merely to cater for but *reinforce* the differences between the different constituencies, and to disguise their common interests; it thereby tends to exacerbate antagonisms between them, in turn justifying a further balkanisation of policy. An example is the community regeneration policies in Oldham, Lancashire, in the 1990s which emphasised the differences between the Asian and white poor rather than the elements of their common plight; this fed into white racism and thus contributed to the racist attacks in 2001 (Beynon and Kushnik, 2002). The 'balanced' approach therefore ends up as a form of divide-and-rule.

What these two modes of dealing with difference have in common is an attempt to stifle real antagonisms within localities formally and bureaucratically. These antagonisms need to be openly contested, precisely so that they can be transcended, and so that common interests can be revealed. But this would go directly against CI's wish to *depoliticise* regeneration through apolitical discourses of 'community', 'fairness', 'diversity' and 'inclusion'.

A similar naivety informs CI's strategy of area improvement through partial gentrification. The eviction of working-class residents by the incomers – that is, their antagonism – is ignored (section 10.4). But the strategy also largely fails as a means of cultural upgrading of the poor. The incoming middle-income residents are spatially separated into distinct enclaves, an ideal condition for investment by speculative builders (Byrne, 2000). For social and cultural reasons, the new residents mix very little with the old, aided in this by the spatial identifier. Using their greater mobility, they consume private services – including schools – outside the neighbourhood (Burchardt *et al.*, 1999) The result is that any existing community organisation of the poor is disrupted by evictions and demoralised by the patronising strategy (Haq, 2004).

Small areas, small achievements

CI's focus on 'the worst areas' has many problems. First, it downgrades the importance of the majority of the poor who do not live in these areas – those in ordinary city neighbourhoods, in 'rich' districts and rural areas. CI gives less priority to these poor because they are not seen to suffer from a collective culture of poverty, and because they do not riot.

Second, targeting the poorest areas and leaving aside areas which are almost as poor actually isolates the former. It tends to further stigmatise them in the eyes of employers and gatekeepers, and it generates resentment from other working-class people in the locality.

Third, targeting small areas fails to address larger scale markets and larger scales of public service provision and regulation in employment, production, housing and education (section 8.5). This can lead to the intervention into the target area being ineffective, for instance the failure of a neighbourhood training scheme because of a job deficit in the sub-region (Hall, 2003). Neighbourhood and local partnerships seldom collaborate with adjacent areas.

Fourth, it is not only the spatial scale but the *time* scale of small-area initiatives which ghettoises them. They are of limited time span, usually three to five years. The assumption is that the spirals of decline can be reversed and virtuous circles established *because* the problems are essentially internal to the area. But the problems are in fact both more deeply rooted internally and more strongly connected to external circumstances that CI supposes, and are seldom resolved by short term initiatives. Projects are therefore closed when they are obviously still needed, and local agencies have to spend time once more attempting to play the funding game.

Fifth, some single-issue zones are intended, in part, as innovative experiments which, if judged successful, may be generalised into the universal service – so-called 'mainstreaming'. But in practice this very seldom occurs. Radical innovations are either not transferred at all, are diluted, or are turned on their head because they are judged *too* empowering and politicising or too expensive. Thus when Labour introduced the Employment Zones they paid volunteers to explore new ways of overcoming barriers to employment and provided a joined-up partnership structure to put those ideas into practice. But this approach was abandoned after six months to be replaced with a centralised, authoritarian, business-run workfare scheme (Jones and Gray, 2001). Similarly, the Sure Start initiatives for young children and their parents were initially concentrated in poor neighbourhoods, provided a wide range of forms of support, and were substantially governed by the parents. In 2005 the programme was main-streamed, but its unit funding was cut and parental control abandoned (Glass, 2005). The *politically* most promising aspects of small-area programmes are thus negated by the government's authoritarian instincts.

These points suggest that small-area programmes tend to ghettoise initiatives and exacerbate social-spatial divisions.

Vicious circles continue to circle

CI's basis in neoliberalism and timidity in the face of business have prevented it from reversing some of the key vicious circles we identified in Part II.

CI seeks to combat **gender disadvantage** among the poor principally through improving women's access to, and gains from, waged work. But these gains are overshadowed by processes which CI does not touch or which it worsens. CI does not force employers to change gender divisions in employment. Nursery provision remains grossly inadequate for the poor, and it is not generally available to children of non-working parents. Poor public services for the poor hit women as carers particularly hard. CI has not addressed either the land-use patterns nor the quality of public transport which severely constrain the lives of poor women. Moreover, CI's priviledging of *waged* work tends to further devalue unwaged caring work – a particular problem in the English-speaking countries (Levitas, 1998). CI cannot conceive of people being socially included *through* caring work, since this would be merely integration into an autonomous working-class community and not into the spaces of business.

Labour has failed directly to address **'racial' divisions** within waged labour. Moreover, many of its policies actually worsen racism (Beynon and Kushnik, 2002). Its enthusiastic participation in the imposition of the USA's new world order (Harvey, 2003) has revived British imperial assumptions including the acceptance of torture of people of colour by the British military and indefinite incarceration without charge of alleged terrorists. Labour's policing of the 'terrorist threat' since 9/11 has produced a sharp increase in police stops and arrests of apparent Arabs or Moslems. Labour's continued immigration controls and clamp-down on refugees further exacerbate popular racism (Hayter, 2004; section 7.3). All this follows logically from CI's embracing of *neoliberal* globalisation and its geopolitical consequences. Domestically, Labour has done nothing substantial to reduce police abuse of black communities. CI's concern for order has ensured the legislation of yet more powers for the policing of the streets and a yet-greater tolerance of police priorities and illegalities.

Regarding crime, the rate of theft has declined under Labour in line with the reduction in the number of those in poverty. But other ways in which neoliberalism has exacerbated crime (section 6.10), including ever-widening income differentials, have not been addressed. CI is if anything *less* sympathetic to the decriminalisation of hard drugs than pure neoliberalism with its potentially libertarian aims. Labour's attempts to instill a different attitude to crime among the poor is contradicted by its own promotion of proprietarian individualism and 'enterprise'. Moreover, the rate of incarceration for given convictions has risen, the prison population increased, and the conditions in prisons and borstals worsened; these follow logically from the authoritarianism of organic-conservatism.

Holistic policy fragmented

CI aspires to joined-up governance for the poor, a comprehensive strategy expressed in a community plan. But this has been contradicted by neoliberal-ism's fragmentation of government (sections 5.4, 5.9 and 9.4). The weakening of the local authority, the greater autonomy for units of delivery of public services, the spinning-off of functions to independent agencies, and the proliferation of partnerships have made serious coordination of policy virtually impossible (Wilkinson and Craig, 2002; Byrne, 2002). Because the problem cannot be solved by re-empowering the local authority, *additional* coordinating agencies have been created by central and local government, starting with the Task Forces in the 1980s. The Blair government has addressed the issue by coordinating poverty policy across ministries through the Cabinet Office and the Government Offices for the Regions, and requiring the formation of Local Strategic Partnerships (LSPs) in each locality. But the LSPs do not have power to direct the spending and policies of even the public sector bodies represented on them. More importantly, there is no bottom-up coordination. If the poor wish to develop holistic strategies they have to attempt to do so through a plethora of bodies not merely within the neighbourhood but at larger spatial scales.

Strongly integrated local strategies, then, tend to be produced only when their authors are limited to the top levels of the local authority and leading business people; but then the coherence of strategy is bought at the expense of democracy, in particular a real input from the poor. A graphic example was Newcastle's 'Going for Growth' strategy, sprung on the city in 2001. While it purported to be a new way of addressing poverty and exclusion, there had been no consultation with poor communities; thousands of people learnt that their homes were to be demolished through press reports (Byrne, 2000).

These *institutional* barriers are in part a reflection of CI's *political* constraint on holism: whatever strategy is adopted, the freedom of business must be respected. While the public, voluntary and community sectors can in principle be coordinated, the actions of business cannot. In particular, there are no mechanisms to control private investment and thus supply jobs. This leaves a gaping hole in any joined-up strategy against exclusion.

Democracy denied

CI's aspirations to improve the poor's access to (local) politics has been severely constrained by spatial political economy and by class biases in the political process. First, local economic, social and regeneration policies, however for-mally democratic and inclusive, are formulated under the **rule of business** constructed by neoliberalism. The continuing privatisation of public services and their quasi-market reforms have further weakened the possibilities for democratic control. These neoliberal constraints are expressed through compe-tition – between localities for investment and jobs, and between units of public services and regeneration bodies for central funding. Local land use planning, for example, is now strongly biased towards developers (Hall and Hubbard, 1996). The options are thus severely limited. The first strategies formulated by the RDAs, set up in the name of regional 'decentralisation' and 'autonomy', focused overwhelmingly on business competitiveness and indeed were virtually identical in the way they conceived this (Painter, 2002).

Second, internal centralisation of power **within the state** has reduced democracy. Control of local government by central government has intensified under Labour. CI's rhetoric of local variability and decentralisation is negated by its neoliberal wish to limit local government spending and to ensure that services and local regeneration initiatives are 'pro-business'. Moreover, Labour has taken further the centralisation of power *within* local authorities onto cabi-nets, chief executives and mayors, removing it from councillors. The logic of these centralisations is that they prevent excessive popular pressures and enable strong but conflicting demands from different groups to be managed.

Nor do **partnerships** fulfill their supposed role of involving the poor. The state still tends to maintain strong control within partnerships: the local authority typically plays a directive role, while the national state maintains tight control of national urban programmes through the criteria it sets, through vetos on projects, and by tying on-going funding to 'output' measurements via relentless monitoring (Stanton, 1996; Hall, 2003). The other major player remains business, as in neoliberal partnerships, with the property sector predictably influential. Business is especially dominant in city-wide and regional bodies, but often does not bother with neighbourhood partnerships (Peck, 1995). The various state and para-state bodies represented are usually protective of their own territory, and much of the policy process consists of competitive maneouvres by these bodies (Allen and Cars, 2001). Partnership has mostly failed to open up local governance to new players: Goodwin (2003) found in west Wales that the same people turned up on the boards of the archipelago of agencies and partnerships, continuing the longstanding dominance of local elites in local politics.

There is, in fact, little political will to open up the political process (Carley and Kirk, 1998: 16). In relation to the general public, the new forms of governance are more opaque than local government. They lack even the limited public debate by elected representatives. Minutes and performance data of the new bodies are often not published and meetings are often closed, sometimes on the grounds of 'commercial confidentiality' – emphasising again the subordination of local democracy to business (Peck, 1993). Democracy is increasingly sidelined to advisory panels with the real decisions taken by 'implementation' boards (Carley et al., 2000: 19).

In consequence, community involvement in partnership bodies is patchy and has not increased over time (Chanan, 2003). The most powerful organisations of ordinary people, the unions, are usually excluded. Labour Party branches are no longer, as they formerly were in *some* working-class localities, sites for political debate and local organisation – due in a large part to the adoption of CI by the party leadership. The most disadvantaged sections of the poor – youth, the homeless, asylum seekers – are scarcely represented. Middle-class communities tend to fare better within partnerships, as they always have in local politics. Community representatives often find that they cannot even keep up with understanding the convoluted rules for bidding nor with reading the voluminous paperwork; grass-roots, innovative community activists find the bureaucratic routines stultifying (Purdue et al., 2000: 44). Attempts at capacity building do not equip disadvantaged populations with the skills to participate effectively, and knowledge is sometimes held back by those in power (Boland, 1999: 220–2). Where they are able formally to participate, community representatives are heavily contrained even in what they can *speak of* by the agendas set by economic competition, the state and the local elites: as in a poker game, they only propose what they know they can get. Thus what is considered 'local knowledge' is often the agenda imposed by outsiders (Cooke and Kothari, 2001). CI, then, has failed to politically involve the poor.

10.6 THE CONTRADICTIONS OF CONSERVATIVE INTERVENTIONISM

These varied failings reflect some fundamental contradictions within the CI project. CI seeks to maintain the rule of markets, class discipline and freedom for capital established by neoliberalism; yet it seeks also to improve the social reproduction of the poor and involve them actively in bettering their situation. It promotes private profit and individualism, while prescribing social responsibility and order for the poor. An integral part of this contradiction is the two different spatial logics within CI: on the one hand neoliberalism operates through economic mobilities, constant instability and fierce cost competition between territories; on the other hand organic conservatism seeks to increase the density of social, economic and political linkages within territories, to make them durable, trusting and committed, and hence to stabilise them. These two logics constantly come into conflict, with the poor nearly always the losers.

We have seen how the freedom and rights of capital collide with the material wellbeing of the poor. The state does not secure jobs for those being pushed from welfare to work; it does not control the movements of capital in ways that could stabilise local economies and counter downward spirals of competition; it does not improve the quality of jobs on offer to the poor except marginally through the low minimum wage. CI's belief in waged work as the panacea for poverty and as the key discipline for the poor leads to a workfare regime which tends to worsen conditions in the secondary labour market. It simultaneously devalues unpaid caring work. Work in the social economy is hemmed-in by market forces, and jobs in the public sector become intensified and are degraded. These low-skilled, casualised and coersive employment relations for the poor contradict CI's promotion of civic and community responsibility (Moody, 1997; Sennett, 1998).

CI wishes to ameliorate the varied forms of disadvantage suffered by oppressed groups, but it does not tackle the discrimination against these groups by employers, nor the biases against them in public services. CI pretends that antagonisms do not exist among the poor communities that it wishes to involve; it does not wish to stimulate self-organisation of oppressed groups. CI does nothing to soften the violent institutions of the state from the military to the police; its conservatism leads to increasingly repressive powers and practices of the legal apparatuses within poor communities and against black people.

CI hopes to re-socialise poor communities through intensive small-area policies. But these are too small to touch some of the key processes, especially the larger scale organisation of capital and the state. The small-area focus legitimates increasingly unequal welfare standards, and tends to exacerbate competition between working-class communities. This *spatial* ghettoisation of anti-poverty intervention is paralleled by its *political* ghettoisation. Policies for the poor are not applied to the rest of the population: community organisation and the social economy are promoted only in poor areas; their consumption is to be social while that of the rest of society is to be increasingly privatised; environmental awareness is good for the poor, but the gross ecological irresponsibility of corporations and ordinary consumers is not challenged. The implication is that the poor are a breed apart, uniquely immoral and unsocialised, uniquely in need of improvement; despite CI's critique of the notion of an underclass (section 1.5), in practice it treats the poor as such. The *commonalities* of the poor with the rest of the population are disguised; trade unions, consequently, are left out of anti-poverty policy. Similarly, the *links* between the poor and the rest of society are disguised: tax credits for the poor, for example, have not been publicised as redistribution, since that would be to politicise the wider issue of income distribution; and they have not been funded, as would be logical, by increasing taxation of the rich. Poor blacks are to be socially included, but migrants and asylum seekers can be demonised for national-political effect – dissonance between political stances at different scales. This spatial and political ghettoisation of poverty policy covers up the fact that the poor are normal, that poverty is normal, and that they are an integral part of the social system (Part II). Expecting the poor to live by different rules and to make no ripples on the rest of society is not merely hypocritical but impractical. This ghettoisation is, however, logical for CI, since it wishes to ameliorate social exclusion while not disturbing the stability of neoliberal class relations among the majority population.

The socialisation and holistic policy sought by CI come up against deregulated business, privatised services, fragmented public services, weakened local authorities, and the proliferating partnerships and agencies, all of them required by neoliberalism's taming of the state. A return to holistic local governance is refused since it could increase popular demands in each area of public policy as the links between them, and the responsibility of the state, became more evident; it would also expose differences in treatment of different localities. The putative role of the poor as responsible local citizens is contradicted by the over-riding constraint of 'competitiveness', the control of central government over local initiatives, the centralisation of local government, and the domination of partnerships by the state, business and the local great and good. These constraints are a constant disincentive to self-organisation by the poor for their own particular interests; where this self-organisation exists, as in the trade unions and some militant community groups, it is excluded from 'partnership'. Both the lack of holism and lack of democracy follow from CI's overriding wish, inherited from earlier neoliberalism, to avoid *aggressive* political mobilisation of any section of the population. Its projects of socialisation and community for the poor are undermined by the greater imperative of political docility.

These failures of social inclusion have changed the way that the term can be understood. 'Social inclusion' could be understood as a kind of equality, the removal of the divides between poor and not-poor. But its use by New Labour has been as equality *of opportunities*, which can of course leave not only substantive inequality but qualitative divides. Worse, CI policy, as we have noted, recreates these divides through its ghettoising policy interventions. Moreover, the failure to remedy income and material inequality leads to 'social inclusion' being pictured mainly in social, political and cultural terms, burying the problem of poverty.

11

ASSOCIATIONALISM
Social democracy from below?

11.1 POSTMODERN SOCIAL DEMOCRACY

Assocationalism, like CI, is a reworking of social democratic themes in opposition to neoliberalism, but with a deeper and more radical critique. Associationalists argue that neoliberalism is dysfunctional for a modern, knowledge-based capitalist economy: this kind of economy requires complex *non-market* relations between economic actors; it needs stable frameworks for productive investment; and it functions best when workers feel involved and valued rather than placed under neoliberalism's disciplinarian sway (Costello *et al.*, 1989; Lipietz, 1992; Hodgson, 1998). Neoliberalism has also exacerbated the social problems of contemporary society by atomising individuals and households, eroding a sense of belonging to communities, and encouraging anomie and selfishness (Bauman, 2000; Putnam, 2000; Beck, 2001). Further, associationalists criticise both neoliberalism and CI for their refusal to make the state accountable to civil society.

Associationalists seek to pursue more forcibly the social elements of CI. Forms of cooperation between economic actors therefore need to be strengthened, particularly at a local level, and cooperative industrial relations fostered. Strong social capital, voluntary organisations and civil society are key, and they must be built from the bottom up, in contrast to the top-down approach of CI. Local organisation is particularly crucial, as in CI, since it is at this scale that participation and networking is easiest to develop. Stronger civil society is a good in itself, offering a sense of belonging and inclusion. It is also the basis for a democratisation of the state, which needs to be decentred and to be put at the service of civil society. We have here a reworking of eighteenth-century liberalism in which civil society is the sphere of liberty, autonomous from state and capital (de Tocqueville, 2003). A vibrant civil society is to be created with plural voices in open public debate.

Social inclusion and overcoming poverty are a natural part of this programme. Networking of the economy is to produce good quality jobs for all, while cooperative relations with *all* workers benefits productivity. Strengthened community organisations can integrate the socially excluded into the political process and ensure that their needs are met. Indeed, local, grass-roots organisations are of particular value to the poor since they are the means for collective self-help, for basic survival (Mcilwaine, 1998). The poor are then enabled to develop self-help and reciprocity, small and micro-enterprises, and stronger community ties. Associationalists here borrow ideas developed in Third World anti-poverty initiatives centred on self-help. Combating social exclusion is thus part of the *mainstream* programme of associationalism rather than the add-on that it is for CI.

Because of its emphasis on locality, associationalists believe that their programme can be implemented through local initiatives and incremental reforms. Indeed, this decentralised strategy is *better* than large-scale social transformation since it allows for experiment and the expression of spatial and social difference – a postmodern stance. Nevertheless, in no developed country has the associationalist model been widely implemented; it accordingly has to be judged mainly on theoretical grounds. During the 1970s and 1980s in Britain some associationalist-type experiments were sponsored by left local authorities (Boddy and Fudge, 1984), but they were brought to a halt by neoliberal pressure and by the abolition of the authorities. Some community politics in the USA have been developed by left activists in ways compatible with associationalism (Miller *et al.*, 1995). In Britain, these local experiments did however influence New Labour, and associationalist ideas have been incorporated into its programme in a watered-down form (Mulgan, 1991; Giddens, 1998; on these overlaps see Chapter 8). Associationalism itself contains many internal debates, which we touch on below.

11.2 SOCIAL THEORY AND THE OPTIMISM OF ASSOCIATIONALISM

Associationalism borrows from varied academic sources: institutionalist and regulationist economics, Weberian distinctions of the social, economic and political realms, post-modern social and cultural theory, and reworked liberal political theory. A number of broad assumptions concerning present-day society tend to be shared, epitomised in what Hall and Jacques (1991) dubbed the 'New Times'.

First, the world economy is seen as in transition from a 'Fordist' model to some kind of 'post-Fordism'. Products are no long standardised but varied and fast changing. The production process tends to be more skilled and to require more autonomy for, and reflexivity from, the worker (Piore and Sabel, 1984; Cooke and Morgan, 1998). This is partly because the economy is increasingly knowledge-based, and, in left accounts, because of demands from workers for more fulfilling work (Wainwright, 1994). Corporations no longer control industries, as smaller firms exploit innovations. The Fordist world-spatial division of labour between high level, skilled work in central regions and routine, standardised production in the periphery is giving way to regional agglomeration of each industry in which the divisions of high/low skill and dominant/subordinate firm are being eroded (Hirst and Zeitlin, 1989; Storper, 1997), even in previously peripheral areas (Storper and Scott, 1992). Expertise developed through cumulative experience produces an ever-changing world division of labour (Sayer and Walker, 1992).

'Fordist' institutions are consequently withering. State intervention to plan industrial development in association with big capital is too crude, large scale, inflexible and authoritarian for the new economy. National unions capable of organising large-scale action and bargaining are being sidelined by local industrial cooperation and by self-employment. Both traditional social democracy and socialist planning are thus obsolete. As a result, waged and household work are changing. The (male) job for life is disappearing; full time work is becoming less common; new forms of work–life balance are emerging. The 'Fordist' nuclear family with male breadwinner and female domestic worker is becoming rare as varied household forms proliferate (cf. section 6.2).

Patterns of social life and culture are similarly diversifying. Consumption of increasingly varied products supports increasingly varied hobbies, fashion choices, lifestyles and identities. Social life has diversified by household type and sexuality, through ethnic diversity, and by locality. But further, social life has become freed from constraint by social group and social structure and is increasingly determined by reflexive individuals, small groups and (virtual) communities (Lash and Urry, 1994; Giddens, 1991). The variety of images in the mass media, but particularly the interactive internet, fosters this democratic cultural politics. In particular, the 'classic' working-class society of pub, Working Men's Club and football for men and the extended family, neighbourhood and bingo for women is breaking up. Working-class people *choose* their culture, reflected in the explosion of lifestyle programmes and magazines, and have become 'aspirational' and thus potentially upwardly mobile. The old solidarities within work and community have been eroded; the 'monolithic working class' is disappearing. It follows that the traditional socialist strategy based on furthering the common interests of the working class nationally and internationally is obsolete.

This optimistic analysis is compatible with various politics. Neoliberals sometimes appeal to it to justify a fragmented economy, social life and culture driven by individual choice, and a greatly reduced state. CI argues that pragmatic networking in economy and society is sometimes needed to achieve real diversity, flexibility and hence choice (Mulgan, 1991; Castells, 1996). Associationalists, however, put a more strongly collective spin on this. Not only the state but capitalist markets too can suppress diversity; the need is for 'neither plan nor market'. Forms of collectivity need to be developed without attempting to recreate the 'monolithic working class': plural, diverse, multiple and overlapping for each individual and group. A thousand hybrids of social and economic organisation should flower, developing reflexively over time and combining non-market with market relations, collectivity with individualism. These can produce varied products, household structures and lifestyles. They are more democratic than neoliberalism or CI since they grow from the bottom

Figure 11.1 Food cooperative in Meadowell, Tyneside, 1990. Associationalists see commercially-run cooperatives as both outcome and means of community building and development of enterprise (source: Steve Conlan).

up. The state should be subordinate to these organisations, political democracy growing out of practical economic and social democracy (Amin and Thrift, 2002).

11.3 SOCIAL INCLUSION FOR ALL

This strategy informs three major areas of policy, each of which addresses the needs of the socially excluded.

A socialised capitalist economy

Economic strategy is strongly focused on strengthening or creating regional industrial clusters, through encouraging links between firms and institutions, arranging for collective services to firms, and attending to labour force reproduction. In these firms and industries, workers are to have a voice and influence based on, and enhancing, their skills (Cooke and Morgan, 1998). Productive links outside the region, however, remain important (Amin and Thrift, 1992; 2002). Ownership may rest with conventional firms or, as market socialist writers urge, with social enterprises that allow more direct control by workers (Sayer, 1995). These relations of production should be more productive than

Taylorist, neo-Fordist or neoliberal models because of better knowledge generation and transfers and because of workers' active involvement (Storper, 1997; Kenney, 1997). Restructuring towards such a model of local industries was pursued by some of the left councils in the 1980s (GLC, 1985; Best, 1990). At a larger scale, associationalists argue that strong, inclusive governance is needed to steer the local economy as a whole, particularly the transition to a 'post-industrial' city (Healey, 1998; Amin, 1999; section 10.4).

Because of this enhanced productiveness, localities will be able to hold their own within global product and financial markets, so the neoliberal opening-up of international flows is to be welcomed rather than feared. Moreover, if labour cooperates with capital to ensure profitability, the discipline of workers by unemployment and tight money will not be required; a national reflationary stance may therefore be possible (Grieve Smith, 2000). Through this strategy, good, stable, fulfilling jobs for all who want them can be provided, eliminating the secondary labour market.

Some associationalists adopt a stronger small-is-beautiful approach, drawing on Proudhon, anarchism and the experience of utopian settlements. Workplaces and enterprises are to be small. The world and national divisions of labour, and their associated trade, financial and ownership flows, are to be opposed: supplies should be local, communities should be as much as possible self-sufficient, and hence demand is kept in the locality. Through this strategy, poor communities can regain real control of local economies devastated by national and global flows. The 'New Urbanism' chimes with this vision (Harvey, 2000: 169–73).

Many associationalists argue that greater economic equality is required, as much for social and political reasons as for material well-being. Without adequate material resources, people will not have the time and means for voluntary work and political participation; evidence is that such engagement is currently much greater among the middle class than the working class and poor. Moreover, large inequalities of income and wealth make it impossible for social trust, and hence social engagement, to grow (Szereter, 2001: 67). Accordingly, measures are needed to redistribute income and wealth. This could be done through the taxation and benefit system; more radically, the state could impose an (initially) equal distribution of economic assets (Roemer, 1996).

From state to governance

The state should become an enabling state, subordinate to civil society. It should foster a dense and inclusive web of business, community and cultural networks and involve these in local governance, creating a 'thick institutionalism' (Amin, 1999). Associationalists argue for more radical forms of decentering of the state than CI, including citizens' juries, alternative dispute mediation, and grass-roots problem-solving, whose open form gives them the potential to involve the poor (Hambledon, 2003). Successful examples include Chicago's neighbourhood governance councils that have taken on powers over policing and education

(Fung, 2003). Partnerships are a desirable form of governance; but to overcome their present lack of democracy (section 10.5) they must solicit and allow *real* participation and agenda setting from community groups, particularly of the socially excluded.

If they are genuinely inclusive, such forms of governance can resolve conflicts within the population and thus create a local consensus. Healey (1998) argues that the key to this is better forms and lines of communication, which can create a Habermasian 'communicative rationality'. A local discursive consensus then puts pressure on all social actors to conform to collective wishes, whatever their private interests. In this way, local governance can *potentially* examine, debate and control all aspects of society, particularly those that create social exclusion; Fung and Wright (2003) term this 'empowered participatory democracy'.

Community, reproduction and social enterprises

Associationalists wish to strengthen community ties. But they generally reject the communitarianism of writers such as Etzioni (1993) who wishes a 'return' to a 'traditional', thus singular, community based around stable families. Associationalists rather seek multiple communities, whose internal social capital is complemented by 'bridging capital' to mutually relate them. Building communities of the poor – based variously around different localities, ethnicities, ages and so on – is perhaps particularly necessary since the poor are seen to lack social capital.

Associationalists wish to build such communities around *practical* organisation to meet economic and social needs; because these needs are most pressing for the poor and socially excluded, this strategy promises to involve them. Practical involvement is most directly achieved through social enterprises. Associationalists have a more radical vision for these than CI (Williams and Windebank, 2002). By providing essential services to poor localities and offering work with a substantial degree of autonomy, relations between producers and consumers, while still partly commoditised, become more direct, thus bridging a gap deeply embedded in capitalist societies. A wide range of basic services could be provided such as credit-union type pensions (Turner and Fichter, 1972). Associationalists aim to use the social economy to keep demand flows within the poor neighbourhood. They also see social enterprise as a way of breaking up and democratising the welfare state. The universal welfare state robbed communities of the mutuality developed in the second half of the nineteenth century; contracting out by the state to social enterprises can return welfare services to neighbourhood control. It can allow imaginative new solutions substantially freed from the routines of professionals and welfare bureaucracies and move across traditional departmental divisions; the writings of Ivan Illich (e.g. 1983) are an inspiration here.

Social enterprises may be one-off visionary projects that combine many elements (Leadbeater, 1997). But associationalists wish to network them through forms such as Community Development Trusts or the cooperative network of Mondragon in Spain; these can strengthen each enterprise by fostering productive exchanges and through cross subsidy. They may achieve viability by using unpaid or training-wage labour, drawing on local goodwill. They may also receive substantial state subsidy and encouragement: in Scotland this has enabled the growth of the social economy over 20 years to comprise 2,700 enterprises employing 25,500 people with value added of £200m per annum (Gordon, 2002). State power may be needed to enable community control. The state may buy land and premises to enable communities to wrest control from the private sector, as in the purchase of expensive land for social housing at Coin Street in London in the 1980s following a community campaign (Brindley *et al.*, 1992: chapter 5). In the Dudley Street neighbourhood of Boston a community coalition was able to make the Redevelopment Agency apply legislation intended for the hotel industry to allow it to buy out slum landlords (Medoff and Sklar, 1994). These examples exemplify the associationalist ideal of the state empowering the community. However necessary the state involvement, associationalists wish the *ownership* of the social economy to be community based. Thus in the Boston case funds were obtained to enable the social housing to be vested with the community. Community ownership can enable the accumulation of capital for investment in new enterprises, as is done by the South Shore Bank in the black ghetto of south Chicago (Taub, 1994), thereby making the sector more autonomous and innovative.

Some associationalists have a more radical vision of the social economy as a comprehensive alternative to modern capitalism, involving various articulations of household self-sufficiency, artisanal work, low technology, low productivity production, and bypassing of commodity exchange. One form is LETS, the local barter of skilled, free-time work (Lee, 1996). Another is deep green agriculture such as permaculture, requiring much labour with low value productivity. Some strategies seek to *combine* this realm of self-realisation and decommodification with work in the mainstream capitalist economy. Thus Fairlie (1996) proposes that everyone should have land for subsistence farming while performing casual work in the capitalist economy. Gorz (1982) has argued for a dual economy comprising self-determined artisanal, non-waged work, paid for by a capitalist economy working as a dictatorship of capital (which he regards as technically inevitable). Others have argued for a citizens' income paid by the state to all, enabling the development of varied forms of individual and cooperative unwaged labour, including caring work.

Even in its less ambitious (or utopian) forms, the associationalist vision of the social economy is radical. We have seen that poverty has many different dimensions aside from income, and varies between different social groups. Social enterprises can address many of these problems – housing, health, care and so on – in a joined-up way tailored to particular sections of the poor. They can provide not only ordinary waged employment but also, more radically, can erode the boundary between work and social life. They may also develop a dynamic of involvement: exceptionally strong participation in the Dudley Street case was developed by designing projects aimed at increasing collective organisation and by keeping residents in control. That gave them the strength to resist cooption by the voluntary sector and local government. With this approach, the social economy promises to be empowering for the socially excluded.

11.4 CRITIQUE OF ASSOCIATIONALISM

There are ambiguities and elisions in the programme of associationalism which point towards difficulties. Is associationalism in the grain of contemporary society or against it? To what extent can islands of associationalism be stably constructed? Are capital and the state major barriers? And, to what extent can the capitalist state become an enabling state for those other than capitalists?

The social analysis to which most associationalists subscribe (section 11.2) is overly optimistic. In Part II we have argued that the empirical patterns it theorises are not mainly due to the coming of an 'informational', 'flexible' or 'reflexive' society but rather to specifically capitalist forms of development, some deepened by neoliberalism, and embodying the power of capital and forms of social oppression. In particular, waged work is not simply developing towards more worker autonomy: intensification of work is found throughout the economy; low skill, Taylorised work flourishes; a large proportion of employment is insecure. This is partly because many goods and services are mass-produced, if with minor variations. Industrial districts with good, stable jobs cannot be created everywhere. The associationalist programme therefore tends to privilege workers in the 'higher' parts of the economy, exacerbating social and spatial inequalities (Gough, 1986). The associationalists' view that social diversity is increasing and is beneficial plays down the sharp differences in *power* (gender, 'race', age, ability) in which this diversity is embedded *and expresses* (Callinicos, 2000). Similarly, the belief that cultural diversity is expanding neglects the power of the media and entertainment industries and, more profoundly, the shaping of culture by capitalist and patriarchal social relations. This is not to say that there is no social foundation for the associationalist project, rather that it has to be created *against* key forms of social power (Cumbers *et al.*, 2003). The crucial question is therefore how countervailing popular power is to be organised.

We have seen that contemporary CI at least formally supports many associationalist policies. Associationalists hope to use these, on a local basis, to further their strategy. Yet we have seen that these projects are severely limited by their neoliberal context, by the autonomy and power of business, and by the subordination of community organisations to the state. Indeed, CI uses community social and economic initiatives to depoliticise poverty. The question is then how these limits can be overcome: how can associationalist initiatives avoid serving neoliberal aims? Again, the question of agency is raised.

We can examine these problems more closely in three fields of associationalist policy:

- economic policy and the mainstream economy;
- the social economy;
- community, social capital and power.

Economic policy and the mainstream economy

The experience of current associationalist-type local economic initiatives in the present period demonstrates the intense pressures that they face from capital and the state. For example, the Enterprise Boards set up by left Labour councils in the 1980s intervened into local industries and enterprises with the combined aims of increasing profitability and growth, combating discrimination in employment, improving wages and conditions, and increasing involvement of the workforce, unions and local communities. But in practice, profitability and employer control tended to override or actually negate the other aims (Harris, 1986). The intervention of the Greater London Enterprise Board to restructure the furniture industry towards a collaborative industrial district fell foul of employers' refusal to collaborate in rationalisations and to allow the union a significant voice (Mackintosh and Wainwright, 1987: chapter 8). The combined influence of board and union was insufficient against the neoliberal habits of business.

These examples raise a deeper issue concerning the associationalist aspiration for 'cooperative' industrial relations: what *concessions* from workers are needed to persuade business to invest and provide jobs? Denmark is a country often cited by associationalists as something of a model of a democratic, negotiated economy (Amin and Thomas, 1996); but in the last 20 years labour has had to make many concessions to business under the threat of disinvestment and high unemployment (Etherington and Jones, 2004). How far do these 'need' to go? Beyond production, how much should the reshaping of the social realm be geared to economic competitiveness: are schools and training providers to teach what employers (think they) want or are they to further students' self-development? Is the benefit system to benefit employers or serve the needs of carers and those who cannot work? Associationalist strategy is silent on these tensions (Zuege, 1999); implicitly it appeals to its optimistic analysis in which workers' interests and capital's profitability are in harmony.

The privileging of the local scale by associationalists exacerbates these problems of power. The smaller the scale of politics, the more intense is the pressure of competitiveness and disinvestment. Citizens are therefore under even greater pressure at the local scale than the national to shape regeneration initiatives to capital's agenda, to see himself or herself as 'in the same boat' as local business (Eisenschitz and Gough, 1997; Gough, 2002). Associationalists lay great store by local cultures and the use of culture in regeneration. But the two dominant modes in which culture is currently being used in local politics are as place marketing through promoting middle-class culture, and as a culture of cooperation between local firms and institutions; cultures of the working class and the poor are ignored or suppressed in the name of local competitiveness (section 10.4). Given their wish to enhance local productiveness, how can associationalists combat this pressure?

Associationalists are also vague about the kind of powers communities might need to pursue their economic projects *effectively*. What kind of economic, social and environmental targets should be set? Should community influence extend to community ownership of all land and the building industry, or community control of the environmental impacts of local enterprises? The failings of existing initiatives (section 10.5) indicate that these are crucial issues.

As we have seen, even 'economically-realist' associationalists seek redistribution of income from rich to poor by the state. But, particularly from the present starting point of neoliberalism and CI, this would require enormous organised pressure from the working class. Associationalists do not explain how this can be achieved with the class fragmentation they support.

The social economy

There is profound tension within associationalism: what is the relation between production governed by capitalist principles *and* non-profitable production in social enterprises, the home and 'free time'. Among the radicals, the proponents of small enterprise and self-sufficiency ignore the way in which their model violates fundamental capitalist dynamics: the ever-extending division of labour, and competitiveness based on ever-larger scale production, integration and circuits of capital. Associationalists supporting the growth of low productivity, artisanal work similarly need to say how this can relate to an economy whose deepest dynamic is increasing labour productivity (Marx, 1970). The radical associationalists must presume one of two possibilities for this cohabitation:

1. The social economy is isolated so that there are minimal flows of commodities, capital and labour across its boundaries. But then its low productivity means a low standard of living, something which may suit the 'downsizing' middle class but which is unattractive to most of the poor. A socialised economy in the present day needs to operate on a large scale.
2. The social economy receives subsidies from the mainstream capitalist economy. But why should business or workers in the mainstream tolerate this drain? Moreover, to the extent that the alternative economy exhibits attractive social relations such as workers' self-management *and*, through subsidy, provides an adequate standard of living, capital will seek to close it down as a political rival.

It does not follow that radical associationalism is impossible, rather that it cannot avoid confronting the mainstream economy. It has to do this in order to create a sufficiently large socialised economy for an adequate level of productivity, and to neutralise the political opposition of capital.

Some associationalists are more sensitive to 'economic realities'. They envisage a majority capitalist sector geared to contemporary rules of competitiveness, coexisting with, and perhaps able to subsidise, a social economy (Hodgson, 1984). But the latter is restricted by both business and the state:

- Social enterprises often find themselves squeezed by input suppliers (landowners, banks) and customers (contractors for components, retailers, the state). To appropriate a higher share of value added in the chain, the social economy has to spread up- and down-stream. Lack of capital often keeps social enterprises in narrow bounds: for this reason the credit unions are presently restricted in the types of financial service they can offer. Moreover, greater worker commitment does not always counterbalance sharper exploitation and better capital resources in the private sector; the quality and autonomy of jobs in the social enterprise are then put in jeopardy (Hayton, 2000). Conversely, successful social enterprises in a capitalist sea often turn themselves into capitalist businesses.
- Social enterprises, which require state funding to survive, are politically under its control. A state whose central agenda is capital accumulation will restrict the political scope of the community sector. Thus community groups that have an overtly radical political agenda find their state funding cut off (Cochrane, 1986). Corporations can lobby the state to restrict the social economy if it encroaches too much on their markets; US banks did this to limit the credit unions. Thatcher's abolition of the GLC and metropolitan countries in 1986 owed much to their high profile, politicising support for radical community groups. The question is once again raised of how to win state policies opposed by capital.

The implication is that even a localised, modest, community-controlled economy has to spread that control to larger spatial scales if it is to flourish, just as the more ambitious, radical schemes would need to do; this once again means *confronting* both capital and the state. But this runs up against the associationalist strategy of exemplary localised projects.

Community, social capital and power

The associationalist enthusiasm for community organisation, social capital and a state subordinated to them does not acknowledge the seriousness of inequalities of social power. This lacuna is within a certain tradition of community studies that sidelines social conflict and the different political stripes of 'communities' (for a critique see Crow and Allan, 1994). Because community is the foundation of associationalist politics, power both *within* and *between* communities is played down (cf. section 10.5). Minority ethnic communities, for example, are treated as if they were homogeneous and their demands non-contradictory, ignoring the power of (small) business and of men within these communities. Iris Young's critique of communitarianism is apposite here: 'community' may be a means for the perpetuation of oppression (and, one might add, economic exploitation: Kakios and van der Velden, 1984). For example, subsidy and support for minority-ethnic owned business, a popular form of community regeneration, may be a poor way of providing jobs to minority-ethnic workers. Moreover, different communities are unequal in power. Middle-class communities, for example, often get the most out of partnership politics; and their activism is often directly precisely at exclusion, as in the creation of autonomous local governments by higher income suburbs in the USA (sections 5.2 and 6.6). This is a problem for a politics which wants *all* communities to engage in politics but which does not acknowledge class power. Thus the community basis of associationalism leads it to perpetuate or even exacerbate differences in social power.

Associationalists also have a naive view of the social capital of the poor: that there is need for *more* of it. Social capital *as such* is the key good, and its utility exists independently of its aims or connections to the larger world. But as we have seen, the poor do not lack social capital: it is rather that it is sometimes directed towards aims which associationalists see as anti-social or unproductive (cf. sections 6.3 and 10.5). Social capital is specific to its *political-economic aims*, and these are predicated on social power. Social capital is not an independent variable that can be increased irrespective of its context.

The same can be said of citizenship and political participation. We have seen that the formal involvement of the poor in partnerships under CI fails in practice since it is contradicted by the political-economic strategy of CI (section 10.5). Civil society is profoundly conditioned by capital and the state (Mcilwaine, 1998; Das, 2004). For the most part this disempowers civil society, and this is true most of all for the socially excluded. The 'enabling' state, which claims to 'involve' civil society, is not class-neutral, and does not open itself to unlimited community power. Mobilising civil society therefore has to take into account the power of capital and the state internalised within it (Petras, 2000). Civil society is not the *deus ex machina* of progressive politics (Meiskins Wood, 1995).

11.5 ASSOCIATIONALISM AND THE GHOST OF CLASS STRUGGLE

The tensions of associationalism in the end boil down to one: its ambiguous attitude to class and other fundamental forms of social power (Zuege, 1999). On the one hand associationalists wish for greater social, economic and political involvement of ordinary people; on the other capitalism is to be the framework. In consequence, associationalism does not register the tensions in its central concepts. Community, social capital and citizenship as such can lend themselves to any kind of politics; if they are pursued as ends in themselves they are likely to lead to the right, given the power of capital and privileged social groups and the current neoliberal context (Gough, 2002). Associationalism seeks empowerment of the poor to achieve greater equality, but its localistic strategy leaves social and spatial uneven development untouched (Gough and Eisenschitz, 1996a). As de Souza Santos writes, 'democratisation of citizenship space is emancipating only as far as it is united with the democratisation of all remaining structural spaces, and citizenship alone is sustainable only to the extent that it spreads beyond the citizenship space' (cited in Baierle, 2002: 311).

Associationalism seeks to abstract good elements of capitalism – skilled, autonomous work, spatially-rooted industrial districts, altruism and mutuality in neighbourhoods, innovation and self-realisation in social enterprises, re-distributive welfare spending, participation in politics – from the nasty, authoritarian features of capital and the state, based on a one-sidedly optimistic reading of the present period. But we have seen that associationalist projects are actually hemmed in by disciplinarian capital and state. This is not to say that the better possibilities of capitalism cannot be fought for; but they pre-cisely have to be *fought for* against capital and the state that tends to serve it, and sometimes against privileged sections of the population. Success against such power cannot be achieved by sections of the socially excluded alone: it requires the maximum unity of all those disadvantaged by the issue at hand. But this goes against associationalism's conception of agency – *particular* socially excluded groups in *individual* localities. This socially- and spatially-fragmented politics cannot push back the constraints of capital and state on associationalism's projects.

In opposition to pure neoliberalism, the logic of strengthening the social aspects of production and reproduction in order to relieve poverty has been championed by various political currents, though in very different ways: extreme right populism, organic conservatism, CI and associationalism. However, none of these escapes from the conflicts of socialisation with markets and the freedom of capital and the conflict of class cooperation with capitalist discipline. Below we consider the socialist project, which seeks to overcome poverty by imposing social logic *over* the freedom of capital.

12

SOCIALIST STRATEGY
FOR INCLUSION

12.1 FROM OBJECT TO SUBJECT OF POLITICS

A socialist strategy begins with the assumption that poverty and social exclusion are not secondary faults of the social system but are *necessarily* produced by capitalism and the social oppressions linked to it. This was also our conclusion from the analysis of Part II. We have also seen that the policies adopted by CI to abolish social exclusion are rendered largely ineffective by the power of capital and privileged social groups and the support given to them by the state; the associationalist strategy, too, fails adequately to address these. Combating social exclusion has therefore to be directed against these forms of domination, and abolishing social exclusion implies their abolition. Seeing the poor as a group with abnormal characteristics who need to be integrated into mainstream society cannot succeed, since it is precisely the mainstream that generates poverty and exclusion. Paolo Freire (1972: 48) expresses this succinctly: the poor 'have always been inside – inside the structure which made them "beings for others"'.

Socialist strategies against exclusion involve dialectics of action, changing social relations and identities, and knowledge formation, which move back and forth between different spatial scales. They combine the following:

1   *struggles against capital and the state's support for it, intertwined with struggles against forms of privilege and oppression within the population.* These aim for social and economic changes which break with exploitation, transcend competition among workers and shake the rights of capitalist property, creating an opening into a non-capitalist society.

2   *building collectivities amongst working-class people within projects for social provision and within social and political organisations, which extend across territories of varied scales.* This weight of numbers is needed to oppose capital and the state successfully. These collectivities can cumulatively construct trust and mutual support. These help to overcome divisions and antagonisms; these also tend to be lessened to the extent that better material well-being is achieved, since this attenuates petty – but presently all-important – competitiveness and rivalries (Foley and Edwards, 1998).

These collectivities – what Das (2004) calls 'the social capital of the working class' – are the embryo of the basic (spatial) relations of a socialist society.

3   *developing a sense of self-esteem among the poor, a sense that their own desires and culture are valid.* We have seen that this is a central problem (sections 6.4 and 7.2). Self-esteem can develop from seeing that their problems are not due to their own failings but rather to structures of power, (1), and through the respect found within collective organisations, (2). Collectivity can engender a sense of power and hence enable the poor to see themselves as active subjects. As Freire urges, the poor can cease to be 'beings for others' and become 'beings for themselves'.

Processes (2) and (3) together constitute radical changes in the culture of the poor; these, however, differ fundamentally from the cultural change proposed by CI: rather than adapting the poor to the work and social disciplines of the given society, socialist culture seeks to adapt the society to the subjectivity and fundamental needs of the poor.

4   *acquiring knowledge, creating new understandings, and diffusing them.* The process of emancipation necessarily involves understanding better how the structures of power work, how taken-for-granted differences within the working class are constructed, and the common interests underlying these differences (Wainwright, 1994). Development of knowledge can unmask the false naturalising of exclusion in common sense, for example the 'self-perpetuation of the underclass'. Fighting the processes of exclusion, (1), generates interest in, and insights into, their nature. Practical collectivity, (2), requires understanding of both differences and commonalities. Self-respect, (3), allows both recognition and critique of one's desires, and appreciation of the extent of one's dependence on others. It is these sorts of knowledge, very different from those privileged in dominant discourses of the 'knowledge society', which are useful for emancipation (Moulaert and Gonzalez, 2003).

This rather grand formulation should not imply that *only* a total social and personal transformation is worth fighting for – a vapid utopianism. On the contrary, every small step along these four axes is useful since it shifts the ground materially and advances confidence and understanding. Indeed, victories, however small, are vital.

This strategy against exclusion begs many questions. *Who* is to comprise its collective agency? *What scales* of material provision and collective organisation are needed? What is the *scope* of change required? How can the enormous *differences* between excluded groups be accommodated and bridged? There are many disagreements among socialists on these questions; in the rest of this chapter we make some contributions to this debate.

Figure 12.1 Demonstration in London against the poll tax, 1990. When the Thatcher government changed local domestic taxation to a flat rate irrespective of income, militant local campaigns started which supported the many people who refused to pay. These coalesced into a campaign across Britain. This played a major role in discrediting Thatcherism and in Thatcher's demise as prime minister, and forced replacement of the poll tax by the council tax (section 5.4) (source: Steve Conlan).

12.2 *WHO* CAN FIGHT EXCLUSION? THE SCALES OF COLLECTIVITY

The strategy just outlined is one of self-emancipation; but of whom? The starting point is the self-activity of the poor and excluded themselves. *Resistance* of the poor to power is ubiquitous and continuous, of course. It is for the most part embodied in the smallest gestures and remarks – the obstinacy of women in the face of their husbands, the bored child being 'stroppy' at school, or the sweatshop worker going slow. Sometimes resistance takes the form of individual heroism, a desperate jump: the single mother who goes 'on the game' to hold her household together, the black man who works alongside white racists, or the gay teenager who runs away to London. All these demonstrate that, however terrible the situation, resistance is possible. But in themselves these do nothing to change the processes of exclusion.

This *has* been done by **collective organisation of the poor**. The most durable, and generally most effective, form of organisation has been of low-paid workers in trade unions. Thus the development of the 'general' unions in late nineteenth-century Britain made a difference not only to low pay and poor conditions but to getting governments to address poverty in general (Charlton, 2000). In poor residential areas during the twentieth century *durable* organisation was mostly within the Labour and Communist Parties and the cooperative movement. But non-party community organisation in poor neighbourhoods has sometimes been militant and successful and thus achieved durability (section 11.3).

The poor's aspirations have occasionally been expressed by non-local and official levels of political parties. Among Labour-controlled councils, Poplar in the 1920s refused to cut unemployment benefit, Clay Cross councillors in 1975 refused to raise housing rents, and in the 1980s Lambeth and Liverpool Councils resisted cutting services. Some large scale organisations of the poor were dependent on left parties: both the agitation for rent controls on Clydeside during the First World War (Damer, 2000) and the rank and file National Unemployed Workers' Movement of the 1920s and 1930s owed much to the Communist Party, while the 1936 Jarrow March was organised by local Labour Party activists (Macintyre, 1980). But the national Labour leadership did not support the 1930s Hunger Marches (Eaden and Renton, 2002: 41–2; Katz, 2001). Higher levels of the Labour Party have taken capital accumulation to be their primary responsibility, and this bias has been magnified by operating through existing state structures (Miliband, 1961). The record of the Labour Party in office and the long term decline of the Communist Party since the 1920s has culminated in the contemporary situation where they now hardly organise the poor even at a neighbourhood level.

However, organisations around single issues of poverty have often grown without strong dependence on left parties. The most significant of these in recent years was the national grass roots campaign against the poll tax; this owed its strength partly to the fact that the whole working class was adversely affected, though the poor most of all (Lavalette and Mooney, 2000). Local campaigns of the poor around housing have long been significant – against council housing rent rises, by squatting, around clearance on central city fringes, and in recent years against transfer of council housing to the voluntary sector. In the 1970s and 1980s there was some local organisation of benefit claimants in Britain; in recent years in France they have constituted a militant national movement, and an EU-wide federation of organisations of the unemployed has organised marches and demonstrations.

The self-organisation of the poor, however, always faces serious obstacles: material obstacles of lack of time and energy and money for meeting, communications and transport; the social obstacles of atomisation and demoralisation (sections 2.3, 4.4, 6.3 and 7.2). Moreover, the non-working poor lack the power to carry out industrial action, the strongest form of power immediately available to ordinary people. The difficulties of effective organisation result on the one hand in lack of any collective action, privatism, often depression, and on the other hand in collective expressions of anger and desperate forms of resistance: vandalising of gentrifiers' houses and cars (as in London Docklands in the 1980s), stealing cars to crash them, fights with the police of white and black youths, riots, or the armed attacks on police of the Black Panthers in the 1960s. These actions are easier to organise than the necessarily long-term and politically complex campaigns to change material conditions; but they are politically ineffective and counter-productive. Riots do sometimes elicit crash programmes in the particular neighbourhood (Barke and Turnbull, 1992); but they do nothing to change national structures of exclusion, simply redirecting palliative state resources.

These weaknesses point to the need to make the struggle of the poor and excluded a part of the struggles of **the working class as a whole** – 'working class' in the Marxist sense of those dependent on wages directly or indirectly. Our analysis in Part II argued that exclusion is produced by *society-wide* forms of power – those of capital, men and whites in particular: the problems of the poor are fundamentally just extremes of those of the working class as a whole. There is thus a basis for common organisation of the excluded and the majority of the population. This gives much greater clout *over particular issues* than the poor alone can muster, because of sheer numbers and because the non-poor have more material resources, greater economic and political influence, and, in some cases, less acceptance of their lot. Even more, if the *fundamental structures* that generate exclusion are to be qualitatively changed, then this can only be achieved by the collective acts of the majority of the population. This is true even of structural variations *within* capitalism: we saw in section 7.4 that different national regimes, which profoundly affect social exclusion, are the product of society-wide class struggle. It is true, a fortiori, of the achievement of a socialist society. This strategy, then, envisages the fight against poverty proceeding through the social movements, the trade unions and movements on social questions, providing that these will attend to the particular needs of the poor.

If we consider social exclusion as multiple forms of oppression as well as poverty, then the **social movements** – of pensioners, women, black people, lesbians and gays, people with various disabilities – have a key role; indeed, their constituencies include the great majority of the socially excluded. It is true that the social movements have significant divisions by income and class, and include conservative wings that are not interested in the problems of the poor. But the *radical* wings of these movements *do* attend to the particular forms of oppression, strongly material as well as ideological, from which the poorer members of these groups suffer – discrimination in employment, poor and biased welfare services, access to housing, harassment by the police – and are thus a key part of organising against social exclusion (Ferguson, 2000; Fletcher and Gaspasin, 2002). A similar conservative/ radical divide exists in **movements on social issues** such as the ecological movement (Pepper, 1993).

The **trade unions** have a particularly important role in combating exclusion because they have the largest membership of any popular organisation and because they can disrupt capital accumulation and material reproduction. For this reason the progressive parts of the social movements often seek union support. The unions organise many low paid and insecure workers, but there are many weaknesses in this protection. Oppressed groups are severely under-represented in union posts. The lowest paid often have low bargaining power and need support from workers with more. The Grunwick dispute in London in 1976–8 showed the possibilities for this (although the struggle failed); but unions are usually unwilling to organise such solidarity – and in contemporary Britain it is illegal. Without large-scale, highly visible union actions that win benefits for low paid workers, the poor will not join unions. Recruitment of workers in the secondary sector is, it is true, made difficult by the very form of exploitation: small workplaces and firms (at the extreme, home-working), lack of permanent contracts, often diverse shifts, and lack of statutory rights arising from size of firm and length of service. Since the 1980s there has been a number of campaigns by voluntary organisations focused on particular residential areas to inform home-workers of their statutory rights. More recently, unions have attempted to recruit in secondary labour markets through 'community unionism': campaigns directed at smaller private-sector workplaces in whatever industry within a particular locality, organised by unions which previously had recruited in only one industry. This is done through both workplaces and local residential areas. The aim is to generate a snowball effect where workers learn that their friends or neigh-bours have joined a union or that the next-door sweatshop has been unionised. A successful example in the last few years has been in the Asian community of west London, where 10,000 people have been unionised using community ties. Community unionism, then, has gone back to an older geography of worker organisation that spans the spaces of production and reproduction (Wills and Simms, 2004). It also opens up the potential for the unions to organise the most exploited workers, those in the informal economy (Gallin, 2001), and to recruit unemployed people in neighbourhoods – if the unions can overcome their traditional reluctance to recruit the unemployed.

Important though the local scale is for unionism of the poor, other scales remain important. National or trans-national firms employ many low paid workers, directly or in franchises. The political influence of unions at a national and EU scale is vital in shaping statutory employment rights. At any rate, effective involvement of the poorest workers depends crucially on political stance. Unions recruit fastest when they are most militant, the second half of the 1960s being a case in point. This means not accepting the logic of inter-firm competition, since competition *proceeds through* the domination of capital over labour (Bryan, 1985; section 12.5 below); this is a key difference between the socialist and associationalist strategies.

The strategy suggested here relies on a strong and confident union movement (on the prognosis in the MDCs, see Frege and Kelly, 2003). Moreover, a greater engagement with the poor can strengthen the unions, not only numerically but also politically and morally. An alliance of the poor with broader working-class organisations can help to radicalise and popularise them. A case in point was the revolt against the poll tax in 1990–1, led by the poor: this began to change the political atmosphere in the country to one where left concerns could once again be voiced.

Popular organisation against exclusion has sometimes spread across the production–residential divide and infused a whole locality (Damer, 2000). In 1969–70 in Turin, for example, there were simultaneously forms of workers' control in Fiat, mass squatting empty flats, and free public transport imposed by its workers. Action in one field inspires others, and the idea can spread: the city *for us*. On a much more modest scale, residents and unionists in Newham in the early 1980s, supported by the Greater London Council, discussed radical plans for all aspects of their locality (Mackintosh and Wainwright, 1987: chapter 10). A corollary of this politics is for unions, community groups and political representatives to move outside their privileged fields of operation (respectively, workplace, residence, state), thus radicalising each.

For these reasons we disagree with Alcock (1997: 201–9) when he argues that the poor, given the barriers to their own organisation, have to rely mainly on the advocacy of professionally staffed voluntary organisations. This neglects the examples of successful organisation by the poor and the potential of social movements and unions.

12.3 POLITICAL FOCUSES, HOLISM AND THE COMPLEXITY OF EXCLUSION

What are the issues around which a socialist approach would build this alliance of the poor and the working class? Harvey (1992) has argued that urban exclusion needs to be tackled on five fronts, namely workplace exploitation, social marginalisation, powerlessness, cultural imperialism and violence. Collective action can start to address these large issues through addressing immediate problems that would make serious inroads into poverty and exclusion while also benefiting the majority of the population. These policies are partly *defensive*, to reverse neoliberal measures, and partly *offensive*, in order to move in egalitarian and democratic directions. For example:

- In *production*, wage differentials need to be narrowed through elimination of low pay and by high level, guaranteed occupational pensions. Casualisation and arrangements designed to fragment labour need reversing in order to improve conditions and strengthen labour's hand. State restriction of trade union activity should be abolished, and stronger workers' rights legislated with serious enforcement that involves the unions.
- *Public services* should be kept in state ownership, with equality and security of units of delivery rather than internal competition and differentiation. A higher proportion of GDP should be spent on these services. On this basis, qualitative improvements can be made to services under the direction of their clients and workers. In some cases such as health for black people and the care of the mentally ill a revolution in the nature of services is needed. Enormous expansion is needed in the funding of some services such as the cultural, leisure and sporting activity of ordinary people.

- All *state benefits* should be at or above the minimum wage, itself set at a decent level, thus linking the respective interests of claimants and workers.
- There should be a switch in *taxation* away from low-income individuals and consumption towards business, the rich and environmental bads – again, with serious enforcement. To illustrate feasible scale: appropriation of a mere 10 per cent of the income of the richest 10 per cent of individuals in Britain could double the income of (or services to) the poorest 10 per cent.
- The multiple forms of *discrimination* in jobs, housing and public services need combating, as part of abating conflicts, competition and exclusions *within* the working class.
- *Social housing* should be maintained in state ownership, though with a large degree of tenant management, giving the potential for (cross-) subsidies and large-scale investment. Investment in it should be increased; the aggressive campaign by tenants on the Walterton estate in inner London suggests one way that this can be done (Rosenberg, 1998). Rent should be no more than a fixed proportion (e.g. 10 per cent) of household income.
- *The public ownership of all land* is in the interests of both the poor and the majority. It would greatly facilitate adequate housing investment and reasonable pricing of it, as well as investment in public and community facilities. It would bestow the role of (potential) developer of all land onto the state, and thus enable *active* land use planning. This is needed to begin to reconfigure urban and rural areas in the interests of accessibility, particularly for women and the poor. Public land ownership would give the state an enormous income stream, facilitating progressive taxation.
- *Public spaces* need *maintaining* as public against, for example, the selling-off of playing fields and parks, the enclosure of streets by retail malls, or appropriation by advertising and corporate sponsorship (Goheen, 1998). But we also need to make public space genuinely open to all. This is partly a matter of combating harassment and violence against oppressed groups by privileged ones. It is also to stop exclusions by police of youths, beggars, street performers, harmless sexual cruisers or political demonstrators. If we add the points above concerning services, housing and land, we have a set of struggles around 'public space' in the widest sense (Mitchell, 2003). These can change perceptions of the public good, ways of living and conviviality, and in particular encourage people to be more open to the socially excluded (Christopherson, 1994).

While they may be pursued through distinct organisations and loci, these policies are not independent of each other. Improvement in one factor is worth achieving – the impact of food on children's attentiveness or the interest rates charged to the poor – but this does not cure social exclusion. This is particularly because of the vicious circles we examined in Part II. For example, the exclusions of women, black people and the elderly are created by the *interaction* of 'economic', 'social' and 'cultural' realms, reinforced by the state (section 7.3); making *serious* inroads into them requires actions on all these fronts. Some vicious circles of exclusion operate over long time spans and are especially difficult to perceive. For example, higher state spending benefiting the poor is desirable, but it may be paid for by higher taxes on working-class people unless taxation is addressed. Similarly, cheaper housing and services are needed, but their effect is substantially cancelled out if employers adjust wages downwards in response and the state reduces benefits. Thus the kind of measures outlined above must be taken forward *as a package* if they are to be effective.

Measures against exclusion need not only to *combine* actions in the economy and the social realm respectively but also to begin to erode the very distinction between them. The ideological chasm between the worlds of work and the home, between employers' and individuals' responsibilities, between toil and 'leisure', is deeply embedded in capitalist society and functions to weaken radical politics of both realms. The quality of poor people's diet, while it has many determinants, cannot be addressed without taking on agribusiness; improving public transport raises issues about the car and oil industries and the economics of car use; and lowering housing costs should throw the spotlight on the house builders and mortgage lenders.

A holistic strategy also needs to address areas of politics not normally considered in social exclusion policy. Combating racist exclusion, for example, cannot be done only at the community and workplace level. The national media play a major role. And 'foreign policy' is crucial, as one can see from the wave of harassment and violence against (supposed) Muslims in the west since the beginning of 'the war on terror'. The anti-war movement is part of the opposition to social exclusion.

Addressing exclusion holistically, as done here, through a class analysis of the society *as a whole* does not mean *reducing* exclusion to one cause. On the contrary, it means taking seriously all social realms since, to use Marx's term, all are 'coloured' by power (Saul, 2002). A corollary is that socialists can usefully reclaim the notion of 'social exclusion'. Rather than its CI meaning which reduces social exclusion to lack of a job, socialists can use it to point to the *internal relations* of 'economic', 'social', 'cultural' and 'political' disadvantage (Ollman, 1993).

A holistic analysis can also make socialists aware of the conflicts that sometimes exist between sections of business and, more deeply, of the dilemmas that business faces between incorporation of labour and discipline over it (sections 1.3 and 4.3; Metzaros, 1970; Harvey, 2003). These are weaknesses which socialist strategy may be able to exploit. For example, manufacturing and commerce may support certain restrictions on finance and landowners to the benefit of the poor, and socialist strategy can exploit the unpopularity of 'parasitic' finance and landowners. More widely, socialist demands can appeal to popular social-democratic ideas on shared class interests which *parts* of business *sometimes* support – economic cooperation, inclusion, political democracy. For example, campaigns for decent training and employment conditions can win popular support by arguing that they ameliorate labour shortages, increase effective demand, and improve productivity. However, the balance in the strategy of business between incorporation and discipline is conjunctural; under neoliberalism it has been largely disciplinary and against measures which benefit the poor. These days, upholding even social-democratic values usually means opposing business.

12.4 THE SOCIAL ECONOMY: SMALL-SCALE AND COMMUNITY SOCIALISATION

The social economy has formed an important element of non-socialist anti-poverty strategies (sections 8.7, 10.3 and 11.1); but it also has promise within socialist politics. It offers to involve people because it is local and visible and addresses immediate needs in a practical way. Because it mostly produces basic necessities, and ones that are relatively undifferentiated, its services can reasonably be provided free to the immediate user, funded by state subsidies. This feature, together with its community basis, facilitates *direct* links between producers and consumers, which can by-pass the commodity form; the building and modification of a house, for example, can be carried out directly to the wishes of those who will inhabit it. In using people's skills and organising abilities within enterprises of their networks, it can develop their confidence in working with others and hence raise their *collective* political ambitions (Wiewel and Gills, 1995). The social economy can thus demonstrate and develop the socialist idea of cooperation to meet common needs.

However, our discussion of the associationalist approach to the social economy suggested that, even in its most ambitious contemporary forms, it is severely constrained: it is constrained by its small scale and lack of internal economies; by business which competes with it or exploits it; and by a state which opposes its potentially anti-capitalist dynamics (section 11.4). Socialists seek to oppose these constraints head-on and thus begin to move the social economy out of its economic, political and spatial ghettos. First, a genuinely social economy should be proclaimed as superior to capitalism and as a model for the whole society. Even associationalism tends to envisage the continuation of capitalism alongside, and usually dominating, the social economy; socialists should have a more aggressive stance. A strong ideology which puts forward the advantages of cooperation, and which is linked to discussing and disseminating practical and political knowledge of it, would greatly strengthen the movement and take it in a socialist direction.

Second, links to workers in the mainstream economy need to be established, in the first place via union membership of social economy workers. Links to public sector workers are needed to ensure that social enterprises are not used to undercut their conditions or employment security (that is, against the neoliberal and CI strategy). Links to workers in private firms, which are competitors or exploiters of social enterprises, can make these problems clear and possibly ameliorate them. Links to all workers can demonstrate to them the advantages of social ownership and workers' management. Third, social enterprises need to be strongly networked not only within localities but also nationally and internationally (O'Gorman, 1995). Fourth, workers and users of social enterprises and their networks should have full, collective control over them at all scales, including ownership of assets. But they need support from the state, both legal-regulatory (status of enterprises; enabling legislation; protection from legal attacks by business) and financial (contracts, purchase of assets, funding of training, subsidies).

With this kind of strategy, the social economy can avoid pressures towards self-exploitation and become genuinely emancipatory. It can then converge with, and enrich, struggles to socialise the mainstream economy.

12.5 DIFFERENT SCALES OF STRUGGLE

Socialist strategy against poverty requires complex articulations of scale. For example, the day-to-day operations of services like primary schools and retailing would appropriately be regulated by workers and consumers on a *neighbourhood or local* scale; but long-term investment in them needs to consider *local or regional* provision and demand (to ensure good provision in poorer neighbourhoods, for example), while some fundamental regulation and funding may need to come from the *national* level. There is a struggle, then to gain control of the service at each of these scales.

In general, necessary scales of control will vary with the particular type of production and consumption involved, technically-determined minimum efficient scales of production, current organisation of capital and the state, forms of uneven development, existing competition between units, spatial flows of economic and political knowledge, and not least, the emerging organisation of workers and citizens. Usually, *multiple* scales will be involved. If we consider an internationally traded sector such as electronics assembly, which employs many disadvantaged workers, a socialist strategy starts in *each factory* around issues of (low) pay, conditions of work, hours and job security. But these cannot be improved on a stable basis without addressing the competition *between* workplaces and shifts of investment *between* them; there is henceforth a need for collaboration between workers across the national and global industry (Gough, 2003a: chapter 13). Similar considerations apply to struggles against transnational companies. In this case the international mobility of capital is easier to see and thus potentially easier to control (Wills, 1998). Their physically harmful products and processes – harming the poor disproportionately – are an international issue, but campaigns often start locally or nationally and then join up (Hensman, 1991). The *global* deregulation of the transnationals through the WTO, which threatens further degradation of working conditions and ecology, has been opposed by unions and other organisations within many localities and countries, loosely coordinated by the movement against neoliberal globalisation (Waterman, 1998; Went, 2000). Popular organisation thus has to be creative in addressing the appropriate scales and – the greatest difficulty – articulating them.

The ability to organise at different spatial scales requires knowledge. This is needed to understand the often hidden links between scalar processes: for example how a *national* workfare programme, inspired by *US* examples, can cause deterioration in a particular *local* labour market (section 5.9). Learning from the successes and failures of political campaigns elsewhere is also crucial. Education and 'popular planning' through union and community organisations have a role here. Organisations involving the poor in the West can learn an enormous amount from union and community struggles in the Third World which are often more inventive and militant (for example, Craig and Mayo, 1995; Barchiesi, 2001).

12.6 INFLUENCING AND CHANGING THE STATE

It follows from the last section that all levels *of the state* are relevant. But what kind of relation can movements against exclusion have to the state in a socialist strategy? We consider two forms.

First, movements can put **pressure on governments** to act in their interest without fundamental change in the state's mode of operation. This occurs partly through electoral and party-political processes, but more importantly through pressure from the movements that can change both policies *and* election results. In most developed countries regional and local governments have sufficient autonomy to make *some* difference to social exclusion, even within the constraints of powers, budgeting and regulation set by central government. Many of the policies outlined in section 12.3 can be adopted. Cleveland city council in the USA, for example, in a hostile neoliberal environment has attempted to 'provide more choices for those who have few' (Krumholz, 1996): it negotiated better provision for low income neighbourhoods with the privately-owned transport provider, and saved the municipal electricity company from a private competitor which was using underhand methods to destroy it. Local councils can give organisational help and ideological support to grass roots campaigning organisations (Wainwright, 1994). To take such policies far, however, means that local councils have to *explicitly* reject the dominant strategy of begging capital to invest and designing all policies to be 'business friendly'; in the absence of sufficient pressure from below, few are at present prepared to take this step.

Contrary to crude notions of 'globalisation' (for example, Ohmae, 1995; Giddens, 1998), nation states, at least of the developed countries, still have the potential power over capital to take actions to ameliorate social exclusion, even some of the radical actions discussed in section 12.3 (Sutcliffe and Glyn, 1999). This is suggested by the very different politics of the state in developed countries, both historically and in the present (section 7.4). Nation states have a wide range of choices for public services and income transfers (Chapter 5); progressive regulation of capital in production and reproduction spheres does not automatically result in catastrophic capital flight. Moreover, the 'rules' of globalisation are substantially determined by the main international agencies, which are directly controlled by the developed-country states. The proposed privatisation of all public services through the General Agreement on Trade in Services, for example, is a political *choice* of these states. Nation states' policies are not a simple response to a given technical-economic environment, but are political choices made under pressures from classes and sections of them, which then determine the 'economic' rules. A sufficiently strong movement of workers and the poor nationally and internationally can substantially affect them (Amoore, 2002; Gill, 2002; Went, 2000).

The impact of working-class movements on the state is, however, severely constrained by its place in capitalist societies. The structures and procedures of the state are far more open to capital than to the working class (Miliband, 1961) and, a fortiori, are systematically biased against the socially excluded. More profoundly, both the state's actions *and* its structures are constrained by the (im)balances of power in civil society (Clarke, 1991), including economic and social exclusions. Accordingly, a longstanding Marxist strategy is to construct a new state consisting of, or closely based on, working class organisations and movements, *in parallel with* these movements achieving socio-economic power (Mandel, 1969). Our second broad strategy, **radical reforms of the functioning of the state within capitalism** can move in this direction. The state can be opened up to partial control by working-class organisations. A weak form of this has been partnerships between local government, community organisations and trade unions, *excluding* business, within which government accepts the policies of the popular organisations. Some of the 'popular planning' done by radical councils in the 1980s was moving towards this method (Mackintosh and Wainwright, 1987). Some of the state reforms proposed by associationalists have this character (section 11.3).

A stronger form is where *major* decisions of the government are explicitly put *wholly* in the hands of popular organisations. A well-known recent example is 'participatory budgeting' (PB) which is now practised, in varied forms, in around 100 cities in Brazil. Most cities are controlled by the Workers' Party, which was faced, until 2003, with a neoliberal federal government. PB has been pioneered since 1989 in Porto Alegre, a city of 1.3 million population with severe problems of poverty. Decisions on priorities for capital investment in the next year are discussed in open neighbourhood assemblies; these send delegates, in proportion to the numbers attending, to 16 'regional' councils which determine the region's priorities; the regions in turn send delegates to a city Budget Council which determines distribution of funds between regions; the city council carries out the Budget Council's recommendations. The obvious importance of these financial decisions, together with the small scales and open forms of the decision-making bodies, have *politicised* local government in Porto Alegre and elicited extraordinary participation from the population, including the socially excluded. In the first five years of the PB, 8 per cent of the adult population was involved at some stage. Involvement was stronger from working-class than middle-class neighbourhoods since the former had more to gain, though the destitute did not participate because their pattern of life did not permit it (Abers, 1998). In the neighbourhood assemblies unskilled workers, those with only primary education, and women have been majorities; black people, subject to informal apartheid in Porto Alegre, have constituted around 20 per cent. The process has stimulated the formation of neighbourhood associations and self-organisation of blacks, disabled people and the elderly (Baierle, 2002). In this way the danger of participation, that it privileges the socially strongest groups (cf. section 10.5), has largely been avoided. Over the years of PB people have learnt how to debate more rationally. There has been a shift from parochial defence of one's patch to support for the most needy neighbourhoods. This exemplifies the general possibilities in socialist strategy for coordination between neighbourhoods and localities, either adjacent ones or ones with similar problems (as in the Outer Estates campaign in Britain). The outcome in Porto Alegre is that investment has switched sharply from large scale, prestigious projects mainly used by the better off to small scale, basic infrastructures such as street paving, sewers and schools. Spending has been more efficient and less corrupt now it is under the supervision of the regional councils (Abers, 1998).

Not surprisingly, there are major problems and tensions in this process. Planning for social needs has been sharply limited by private capital, particularly private land ownership. The *spatial scales* of the PB have also weakened it. Planning of infrastructures *across* the city is weak, resulting in severe malfunctions. The council does not supply data essential for citywide planning, and hence control remains substantially in the hands of the politicians and state bureaucracy with their susceptibility to corporate pressures. The PB process, however, has not been taken to all state governments (Baierle, 2002) nor, since 2003, to the federal level. These problems again point to the importance of linking community to larger scales of control. But the problems of PB are ultimately ones of weakness in popular mobilisations. In Brazil, the trade unions, which in the 1980s destroyed the military dictatorship, were in retreat in the 1990s. Poverty has worsened over the last 15 years because of a neoliberal offensive. In Porto Alegre in the 1990s unemployment soared (Baierle, 2002). Radically reforming the state requires strong popular movements in the social and economic arenas.

12.7 AN INCLUSIVE SOCIALIST SOCIETY

The strategy described above does not in itself go beyond capitalist society. But it points in the *direction* of a socialist society by developing collective organisations including oppressed groups and the poor, by confronting class and other forms of social power, and, perhaps, by achieving some egalitarian and democratic reforms; these are *foundations* for constructing a socialist society. Socialists argue that the latter is necessary in order to overcome the main processes producing social exclusion and realise the full potential of the policies considered above. Having some notion of what a socialist society might look like is useful in developing struggles in that direction (for a vision which overlaps with ours, see Harvey, 2000: appendix).

Socialists envisage major economic assets – fixed capital, economically useful knowledge, investment money – being publicly owned and controlled. In our view the major form of economic planning should be of *investment* rather than of prices or inter-enterprise exchanges. This planning could take place in both territories of varied scales and in individual sectors, mutually articulated (Devine, 1988; Itoh, 1995; Gough and Eisenschitz, 1998). This would enable poverty to be overcome. Waged employment could be available for all who seek it by matching investment rates and productive capacity to the (would-be) labour force. Planning the run down of workplaces and providing for replacement jobs would achieve increased job security. Given the total territorial product, a wages and state-incomes policy can allocate this product according to social criteria (constrained by any labour shortages); this allows a far more egalitarian income distribution than at present.

The implications of labour processes for both workers and ecology could be planned at the enterprise and sector level. Subject to procedures to ensure sufficient, and roughly equal, effort by workers across the economy, the pace, tasks, level of skill, types of sociability and healthiness of jobs could be negotiated; this could abolish monotonous, unskilled, unhealthy and dictatorial jobs (though not always such tasks), benefiting the majority of the workforce but particularly the socially excluded. Targets for the division of labour by gender, 'race', age and disability could be pursued through substantial measures rather than formal affirmative action. Process and product knowledge would be a commons, its spread no longer constrained by capitalist property – the example of Linux software is an inspiration here (Moody, 2001).

Such an economy could also combat the processes of social exclusion in the reproduction sphere. Social ownership of assets and egalitarian distribution principles could enable a higher proportion of GDP to be devoted to social services. These services would no longer be shaped by capitalist–disciplinarian aims (Chapter 5). Through sectoral and enterprise planning, the design and marketing of consumer commodities could be freed from the waste, fraud, planned obsolescence and unhealthiness of capitalist practice (Chapter 6), and designed to meet and develop real needs (Cooley, 1987). In the case of relatively undifferentiated products, consumers could have a direct input (rather than via the market) into the industry's plans. Polluting processes and products could be rejected on a prudential basis, with the whole of society, rather than the enterprise, carrying the opportunity cost.

As part of these processes, workers would have a far higher degree of control over their work than at present, particularly compared with contemporary poor jobs. Subject to the goals chosen by the whole society, they could have active roles, as individuals and in enterprise, sector and territorial collectives, in shaping their jobs and industries' processes, products and ecology (Blazyca, 1983). For most of today's socially excluded, this would constitute a qualitative jump in power, self-development and (to that extent) self-respect.

The possibilities for household formation, caring arrangements, unpaid work and neighbourly work could multiply and blossom, materially supported by increased incomes for the formerly poor, improved social services and good social housing (Harvey, 2000). Many kinds of arrangement of collective unpaid work could be possible in the care of children and the infirm, in cooking, and in small enterprises producing free services. Buildings and durables could be shared in new ways. On this basis, the layout of houses, of streets and blocks, and of the city itself could be radically reconceived (Hayden, 1980; Greed, 1994: 174–84). Together with the changes in waged work, this would allow gender divisions of labour in unpaid work, and hence gender itself, to be eroded (Barrett and McIntosh, 1991; Hennessy and Ingraham, 1997).

Socialist relations in production *and* reproduction would tend to generate a greater respect for both public property (as a collective good) and individual property (as deserved), thus reducing theft and fraud. More generally, material security improves trust between people (Foley and Edwards, 1998) and thus helps, cumulatively, to develop collective organisation and concrete forms of cooperation.

All this has implications for inclusion in public spaces. Socialist forms of control of the waged production of the built environment, public transport, retailing and consumer services could allow public spaces to be more open to diverse activities and social groups. Greater collectivity in the residential sphere implies a stronger collective appropriation of neighbourhood public spaces, and indeed a certain blurring between the 'private' home and the 'public' neighbourhood – the block's nursery, kitchen/dining room and swimming pool, for example. Together, these would generate greater respect for public spaces *and* for people using them.

These relations do not automatically remove exclusion by gender, sexuality, racism, age and ability. But the brief account here does suggest that the social-isation of production and reproduction is *necessary* to inclusion and are a *fruitful base* for these struggles to be waged (Fraser, 1995; Saul, 2002).

The broadest and strongest objection to socialism is that people lack sufficient altruism to make it possible. This would be a particularly serious objection to socialism *as* a way of removing poverty, inequality and exclusion: selfishness and domination will out. Our reply is that a socialist society will not appear fully-fledged, but grow out of the struggles and organisation that historically create it. These are sites for increasing recognition of the claims of others across social and spatial divides, and for recognising one's own real dependence on those others. They are thus the basis for developing social inclusion simultaneously as consciousness and practice.

BIBLIOGRAPHY

Abercrombie, N., Warde, A., Soothill, K., Urry, J. and Walby, S. (2000) *Contemporary British Society*, Cambridge: Polity Press.

Abers, R. (1998) 'Learning democratic practice: distributing government resources through popular participation in Porto Alegre, Brazil', in M. Douglas and J. Friedmann (eds) *Cities for Citizens*, New York: Wiley.

Abrams, P. and McCulloch, A. (1976) *Communes, Sociology and Society*, Cambridge: Cambridge University Press.

Aglietta, M. (1979) *A Theory of Capitalist Regulation,* London: Verso.

Ainley, P. (1988) *From School to Youth Training Scheme*, Milton Keynes: Open University Press.

Albo, G. (1993) 'Competitive austerity and the impasses of capitalist employment policy', in R. Miliband and L. Panitch (eds) *Socialist Register 1994*, London: Merlin.

Alcock, P. (1997) *Understanding Poverty*, 2nd edn, Basingstoke: Macmillan.

Aldridge, A. (1998) 'Habitus and cultural capital in the field of personal finance', *Sociological Review*, 46 (1):1–23.

Aldridge, T., Lee, R., Leyshon, A., Thrift, N., Tooke, J. and Williams, C. (2001) *Bridges into Work? An evaluation of Local Exchange Trading Schemes*, Bristol: Policy Press.

Allen, J. and Cars, G. (2001) 'Multiculturalism and governing neighbourhoods', *Urban Studies,* 38 (12): 2195–209.

Allen, J. and Hamnett, C. (eds) (1992) *Housing and Labour Markets,* London: Allen and Unwin.

Allen, J., Massey, D. and Cochrane, A. (1997) *Rethinking the Region: spaces of neo-liberalism*, London: Routledge.

Ambrose, P. (1994), *Urban Process and Power*, London: Routledge.

Amin, A. (1999) 'An institutional perspective on regional economic development', *International Journal of Urban and Regional Research,* 23 (2): 363–78.

Amin, A. and Thomas, D. (1996) 'The negotiated economy: state and civic institutions in Denmark', *Economy and Society*, 25 (2): 255–81.

Amin, A. and Thrift, N. (1992) 'Neo-Marshallian nodes in global networks', *International Journal of Urban and Regional Research*, 16: 571–87.

Amin, A. and Thrift, N. (2002) *Cities: re-imagining the urban*, Cambridge: Polity Press.

Amin, A., Cameron, A. and Hudson, R. (2002) *Placing the Social Economy*, London: Routledge.

Amoore, L. (2002) *Globalisation Contested: an international political economy of work*, Manchester: Manchester University Press.

Anderson, J. and Jensen, P. (eds) (2002) *Changing Labour Markets, Welfare Policies and Citizenship*, Bristol: Policy Press.

Armstrong, P., Glyn, A. and Harrison, J. (1991) *Capitalism since 1945*, Oxford: Blackwell.

Atkinson, R. (1999) 'Urban social exclusion in the EU: concepts, causes and prospects', *Urban Futures*, 18 (3): 76–92.

Baierle, S. (2002) 'The Porto Alegre Thermidor? Brazil's "participatory budget" at the crossroads', in L. Panitch and C. Leys (eds) *Socialist Register 2003*, London; Merlin.

Baker, J. (1987) *Arguing for Equality,* London: Verso.

Bamber, G. and Lansbury, R. (1998) *International and Comparative Employment Relations*, 3rd edn, London: Sage.

Banks, S. and Shenton, F. (2001) 'Regenerating neighbourhoods: a critical look at the role of community capacity building', *Local Economy,* 16 (4): 286–98.

Barchiesi, F. (2001) 'Transnational capital, urban globalisation and cross-border solidarity: the case of South African municipal workers', in P. Waterman and J. Wills (eds) *Place, Space and the New Labour Internationalisms*, Oxford: Blackwell.

Barke, M. and Turnbull, G. (1992) *Meadowell: the biography of an estate with problems*, Aldershot: Avebury.

Barrett, M. (1980) *Women's Oppression Today,* London: Verso.

Barrett, M. and McIntosh, M. (1991) *The Anti-Social Family*, 2nd edn, London: Verso.

Bauman, Z. (1998) *Work, Consumerism and the New Poor*, Milton Keynes: Open University Press.

Bauman, Z. (2000) *The Individualised Society*, Malden, MA: Polity Press.

Beatty, C. and Fothergill, S. (1998) 'Registered and hidden unemployment in the UK coalfields', in P. Lawless, R. Martin and S. Hardy (eds) *Unemployment and Social Exclusion*, London: Jessica Kingsley.

Beck, U. (1992) *Risk Society*, London: Sage.

Beck, U. (2001) 'Living your own life in a runaway world: individualisation, globalisation and politics', in W. Hutton and A. Giddens (eds) *On the Edge: living with global capitalism*, London: Vintage.

Becker, S. (2003) '"Security for those who cannot": Labour's neglected welfare principle', in J. Millar (ed.) *Understanding Social Security*, Bristol: Polity Press.

Bensham Settlement (2003) *Guide*, Gateshead: Gateshead City Council.

Best, M. (1990) *The New Competition*, Cambridge: Polity Press.

Betz, H.-G. (2002) 'Xenophobia, identity politics and exclusionary populism in western Europe', in L. Panitch, and C. Leys (eds) *Socialist Register 2003*, London: Merlin.

Beynon, H. and Kushnik, L. (2002) 'Cool Britannia or Cruel Britannia? Racism and New Labour', in L. Panitch, and C. Leys (eds) *Socialist Register 2003*, London: Merlin.

Beynon, H., Hudson, R., Lewis, J., Sadler, D. and Townsend, A. (1989) '"It's all falling apart here": coming to terms with the future in Teesside', in P. Cooke (ed.) *Localities,* London: Unwin Hyman.

Bianchini, F. and Parkinson, M. (eds) (1993) *Cultural Policy and Urban Regeneration: the Western European experience,* Manchester: Manchester University Press.

Birkin, M., Clarke, G. and Clarke, M. (2002) *Retail Geography and Intelligent Network Planning,* Chichester: Wiley.

Blackburn, R. (1999) 'The new collectivism: pension reform, grey capitalism and complex socialism', *New Left Review*, 233: 3–65.

Blazyca, G. (1983) *Planning is Good for You: the case for popular control,* London: Pluto.

Boddy, M. (1980) *The Building Societies,* Basingstoke: Macmillan.

Boddy, M. and Fudge, C. (eds) (1984) *Local Socialism?,* Basingstoke: Macmillan.

Boland, P. (1999) '"Community: economic development in Knowsley, Merseyside: rhetoric versus reality', *Local Economy,* 14 (3): 214–31.

Bonefeld, W. (2001) 'European monetary union: ideology and class', in W. Bonefeld (ed.) *The Politics of Europe,* Basingstoke: Palgrave.

Bonefeld, W., Brown, A. and Burnham, P. (1995) *A Major Crisis? The politics of economic policy in Britain in the 1990s,* Aldershot: Dartford.

Booth, W. (1890) *In Darkest England and the Way Out,* London: Salvation Army.

Bourdieu, P. (1986) 'The forms of capital' in I. Richardson (ed.) *Handbook of Theory and Research for the Sociology of Education,* New York: Greenwood Press.

Bourgois, P. (2002) *In Search of Respect: selling crack in El Barrio,* Cambridge: Cambridge University Press.

Bowles, S. and Edwardes, R. (1993) *Understanding Capitalism,* New York: HarperCollins.

Bowles, S. and Gintis, H. (1976) *Schooling in Capitalist America,* New York: Basic Books.

Bowling, A., Grundy, E. and Farquhar, M. (1997) *Living Well into Old Age,* London: Age Concern.

Bowling, B. and Philips, C. (2002) *Racism, Crime and Justice,* Harlow: Longman.

Brandes, S. (1970) *American Welfare Capitalism, 1880–1940,* Chicago, IL: University of Chicago Press.

Brenner, N. (2001) 'The limits to scale? Methodological reflections on scalar structuration', *Progress in Human Geography,* 25 (4): 591–614.

Brenner, N. and Theodore, N. (2002) 'Cities and the geographies of "actually existing neoliberalism"', *Antipode,* 34 (3): 349–79.

Brewer, M., Clark, T. and Goodman, A. (2003) 'What really happened to child poverty in the UK under Labour's first term?', *The Economic Journal,* 113: 240–57.

Brewer, M., Goodman, A., Shaw, J. and Shephard, A. (2005) *Poverty and Inequality in Britain: 2005,* London: Institute for Fiscal Studies.

Brindley, T., Rydin, Y. and Stoker, G. (1992) *Remaking Planning,* London: Routledge.

Broad, D. (1995) 'Globalization and the casual labor problem: history and prospects', *Social Justice,* 22 (3): 67–87.

Bryan, R. (1985) 'Monopoly in the Marxist Method', *Capital and Class,* 26: 72–92.

Bryson, A. (2003) 'From welfare to workfare', in J. Millar (ed.) *Understanding Social Security,* Bristol: Polity Press.

Bryson, J., Daniels, P. and Warf, B. (2004) *Service Worlds,* London: Routledge.

Bryson, L. (1992) *Welfare and the State,* Basingstoke: Macmillan.

Burawoy, M. (1985) *The Politics of Production,* London: Verso.

Burchardt, T. (2003), *Being and Becoming: social exclusion and the onset of disability,* CASE Report 21, London: LSE.

Burchardt, T., Hills, J. and Propper, C. (1999) *Private Welfare and Public Policy,* York: Joseph Rowntree Foundation.

Burchardt, T., Le Grand, J. and Piachaud, D. (2002) 'Degrees of exclusion: developing a dynamic, multidimensional measure', in J. Hills, J. Le Grand and D. Piachaud (eds) *Understanding Social Exclusion,* Oxford: Oxford University Press.

Burns, D. (1994) *The Politics of Decentralisation,* Basingstoke: Macmillan.

Burns, W. (1963) *New Towns for Old,* London: Leonard Hill.

Burrows, R. (2003) *Poverty and Home Ownership in Contemporary Britain,* York: Policy Press/Joseph Rowntree Foundation.

Butler, S. (1998) *Access Denied: the exclusion of people in need from social housing*, London: Routledge/Shelter.

Byrne, D. (1999) *Social Exclusion and the City*, Milton Keynes: Open University Press.

Byrne, D. (2000) 'Newcastle's going for growth: governance and planning in a post-industrial society', *Northern Economic Review*, 30: 3–16.

Byrne, D. (2002) 'Public–private partnerships', *Local Economy*, 17: 335–7.

Byrne, D. and Wharton, C. (2004) 'Loft living – Bombay calling: culture, work and everyday life on post-industrial Tyneside', *Capital and Class*, 84: 191–8.

Cabinet Office (2000) *Minority Ethnic Issues in Social Exclusion and Neighbourhood Renewal*, London: Social Exclusion Unit.

Cabinet Office (2002) *Social Capital: a discussion paper*, London: Performance and Innovation Unit.

Callinicos, A. (2000) *Equality*, Cambridge: Polity Press.

Campbell, B. (1993) *Goliath: Britain's dangerous places*, London: Methuen.

Campbell, B. (1999) 'Masculinity, poverty and crime', in J. Vail, J. Wheelock and M. Hill (eds) *Insecure Times*, London: Routledge.

Cant, B. (1997) *Invented Identities? Lesbians and gays talk about migration*, London: Cassell.

Carley, M. (2000) 'Urban partnerships, governance and the regeneration of Britain's cities', *International Planning Studies*, 5 (3): 273–97.

Carley, M. and Kirk, K. (1998) *Sustainable by 2020: a strategic approach to urban regeneration for Britain's cities*, York: Joseph Rowntree Foundation.

Carley, M., and Chapman, M., Hastings, A., Kirk, K. and Young, R. (2000) *Urban Regeneration through Partnerships*, Bristol: Policy Press.

Carmel, E. and Papodopoulos, T. (2003) 'The new governance of social security in Britain', in J. Millar (ed.) *Understanding Social Security*, Bristol: Polity Press.

Carrier, J. and Kendall, I. (1998) *Health and the NHS*, London: Athlone Press.

Cars, G., Madanipour, A. and Allen, J. (1998) 'Social exclusion in European cities', in Cars, G., Madanipous, A. and Allen, J. (eds) *Social Exclusion in European Cities*, London: Jessica Kingsley.

Castells, M. (1982) *City, Class and Power*, Basingstoke: Macmillan.

Castells, M. (1996) *The Rise of the Network Society*, Oxford: Blackwell.

Cattell, V. and Evans, M. (1999) *Neighbourhood images in East London*, York: Joseph Rowntree Foundation.

Centre for Environmental Studies (CES) (1985) *Outer Estates in Britain: a framework for action*, London: CES.

Chanan, G. (2003) *Searching for Solid Foundations: community involvement and urban policy*, London: Office of the Deputy Prime Minister.

Charlton, J. (2000) 'Class struggle and the origins of state welfare reform', in M. Lavalette and G. Mooney (eds) *Class Struggle and Social Welfare*, London: Routledge.

Chatterton, P. and Hollands, R. (2003) *Urban Nightscapes: youth culture, pleasure spaces and corporate power*, London: Routledge.

Chauncey, G. (1994) *Gay New York: gender, urban culture, and the making of the gay male world 1890–1940*, New York: Basic Books.

Cheshire, P. and Sheppard, S. (2004) 'Capitalising the value of free school: the impact of supply characteristics and uncertainty', *Economic Journal*, 114 (11): 397–424.

Christopherson, S. (1994) 'The fortress city: privatized spaces, consumer citizenship', in A. Amin (ed.) *Postfordism: a reader*, Oxford: Blackwell.

Clark, D. (2004) 'Unto him that hath', *The Guardian*, 6 August.

Clarke, S. (1988) *Keynesianism, Monetarism and the Crisis of the State*, Aldershot: Edward Elgar.

Clarke, S. (1991) 'State, class struggle, and the reproduction of capital', in S. Clarke (ed.) *The State Debate*, Basingstoke: Macmillan.

Clarke, S. (1992) 'What in the F. . . .'s name is Fordism?', in N. Gilbert, R. Burrows and A. Pollert (eds) *Fordism and Flexibility*, Basingstoke: Macmillan.

Cloke, P. and Little, J. (eds) (1997) *Contested Countryside Cultures,* London: Routledge.

Cloke, P., Milbourne, P. and Widdowfield, R. (2002) *Rural Homelessness*, Bristol: Policy Press.

Coates, D. (2000) *Models of Capitalism*, Cambridge: Polity Press.

Cochrane, A. (1986) 'Community politics and democracy', in D. Held and C. Pollit (eds) *New Forms of Democracy*, London: Sage.

Cockburn, C. (1983) *Brothers: male dominance and technological change,* London: Pluto Press.

Coffield, F., Borrill, C. and Marshall, S. (1986) *Growing up at the Margins,* Milton Keynes: Open University Press.

Cohen, R. (1991) *Contested Domains: debates in international labour studies,* London: Zed Books.

Cohen, S. (1985) *Visions of Social Control*, Cambridge: Polity Press.

Cole, I. and Furbey, R. (1994) *The Eclipse of Council Housing*, London: Taylor & Francis.

Cole, K., Cameron J. and Edwards, C. (1983) *Why Economists Disagree*, Harlow: Longman.

Colenutt, B. and Tansley, S. (1990) *Inner City Regeneration,* Manchester: Centre for Local Economic Studies.

Cooke, B. and Kothari, U. (2001) 'The case for participation as tyranny', in B. Cooke, and U. Kothari (eds) *Participation: the new tyranny?*, London: Zed Books.

Cooke, P. and Morgan, K. (1998) *The Associational Economy*, Oxford: Oxford University Press.

Cooley, M. (1987) *Architect or Bee? The human price of technology*, London: Hogarth.

Coontz, S. (1988) *The Origins of Private Life: a history of American families 1600–1900,* London: Verso.

Costello, N., Mitchie, J. and Milne, S. (1989) *Beyond the Casino Economy: planning for the 1990s,* London: Verso.

Cox, D. (1998) 'Australia', in J. Dixon and D. Macarov (eds) *Poverty: a persistent global reality*, London: Routledge.

Cox, K. (1989) 'The politics of turf and the question of class', in J. Wolch and M. Dear (eds) *The Power of Geography,* Boston, MA: Unwin Hyman.

Cox, K. (1993) 'The local and the global in the new urban politics: a critical view', *Environment and Planning D: Society and Space*, 11: 433–48.

Cox, K. (1997) 'Globalisation and geographies of workers' struggles in the late twentieth century', in R. Lee and J. Wills (eds) *Geographies of Economies*, London: Arnold.

Cox, K. (1998) 'Spaces of dependence, spaces of engagement and the politics of scale; or, looking for local politics', *Political Geography*, 17 (1): 1–24.

Cox, K. (2004) 'Globalisation, the class relation and democracy', *GeoJournal*, 60: 31–41.

Craig, G. and Mayo, M. (eds) (1995) *Community Empowerment*, London: Zed Books.

Croall, H. (1998) *Crime and Society in Britain*, Harlow: Longman.

Crompton, L. (1985) *Byron and Greek Love*, London: Faber.

Crompton, R. (1997) *Class and Stratification*, Cambridge: Polity Press.

Crouch, C. and Streeck, W. (eds) (1997) *Political Economy of Modern Capitalism*, London: Sage.

Crow, G. and Allan, G. (1994) *Community Life: an introduction to local social relationships,* Hemel Hempstead: Harvester Wheatsheaf.

Culpitt, I. (1999) *Risk and Social Policy*, London: Sage.

Cumbers, A., MacKinnon, D. and McMaster, R. (2003) 'Institutions, power and space: assessing the limits to institutionalism in economic geography', *European Urban and Regional Studies*, 10 (4): 325–42.

Currie, E. (1996) 'Social crime prevention strategies in a market society', in J. Muncie, E. McLaughlin and M. Langan (eds) *Criminological Perspectives: a reader*, London: Sage.

Curtis, L. (1984) *Nothing But The Same Old Story*, London: Information on Ireland.

Damer, S. (2000) '"The Clyde Rent War!": the Clydebank rent strike of the 1920s', in M. Lavalette and G. Mooney (eds) *Class Struggle and Social Welfare*, London: Routledge.

Danziger, N. (1997) *Danziger's Britain: a journey to the edge,* London: Flamingo.

Das, R. (2004) 'Social capital and the poverty of wage labourers: problems with social capital theory', *Transactions of the Institute of British Geographers*, 29 (1): 27–45.

Davies, J. (1972) *The Evangelistic Bureaucrat,* London: Tavistock.

Davin, A. (1978) 'Imperialism and motherhood', *History Workshop*, 5: 9–65.

Davis, M. (1986) *Prisoners of the American Dream,* London: Verso.

Davis, M. (1992) *City of Quartz*, London: Verso.

Davis, M. (2001) *Late Victorian Holocausts*, London: Verso.

Davis, N. (1999) 'Crisis, crisis, crisis: the state of our schools', *The Guardian*, 14 September.

Deacon, A. (1999) 'Learning from the US? The influence of American ideas upon "New Labour" thinking on welfare reform', *Policy and Politics*, 28: 5–18.

Deacon, B. (2002) 'Globalization and the challenge for social security', in R. Sigg and C. Behrendt (eds) *Social Security in the Global Village*, New Brunswick, NJ: Transaction Press.

Deakin, S. and Wilkinson, F. (1991) 'Social policy and economic efficiency: the deregulation of the labour market in Britain', *Critical Social Policy*, 11 (3): 40–61.

Dean, M. (1995) 'Governing the unemployed self in an active society', *Economy and Society*, 24: 559–83.

Dean, M. and Taylor-Gooby, P. (1992) *Dependency Culture: the explosion of a myth*, Hemel Hempstead: Harvester Wheatsheaf.

Dear, M. and Scott, A. (eds) (1981) *Urbanisation and Urban Planning in Capitalist Society,* London: Methuen.

de Brunhof, S. (1978) *The State, Capital and Economic Policy*, London: Pluto Press.

Delanty, G. (2000) *Citizenship in a Global Age*, Milton Keynes: Open University Press.

Department of Social Security (1998) *A New Contract for Welfare: partnerships in pensions*, London: Sationery Office.

Department for Transport (2003) *Transport Statistics: Great Britain,* London: HMSO.

Department of Work and Pensions (2002) *Departmental Report on Government's Expenditure Plans 2002–3*, London: HMSO.

de Tocqueville, A. (2003) *American Institutions and their Influence,* Indypublishers.com

Devine, P. (1988) *Democracy and Economic Planning*, Boulder, CO: Westview Press.

Dirven, H.-J., Fouarge, D. and Muffels, R. (1998) 'Netherlands', in J. Dixon and D. Macarov (eds) *Poverty: a persistent global reality*, London: Routledge.

Dixon, J. and Hyde, M. (eds) (2001) *The Marketization of Social Security*, Westport, CT: Forum Books.

Dixon, J. and Macarov, D. (eds) (1998) *Poverty: a persistent global reality*, London: Routledge.

Dobash, R. and Dobash, R. (1983) *Violence against Wives: the case against patriarchy*, Basingstoke: Macmillan.

Dobbs, L., Moore, C., Craddock, C., Heyman, A. and Driver, H. (2004) *On the Move: women, transport and the labour market in the north east of England*, Newcastle upon Tyne: Centre for Public Policy.

Donzelot, J. (1981) *Policing of Families: welfare versus the state*, London: HarperCollins.

Doogan, K. (2001) 'Insecurity and long-term employment', *Work, Employment and Society*, 15 (3): 419–41.

Dore, R. (1997) 'The distinctiveness of Japan', in C. Crouch and W. Streeck (eds) *Political Economy of Modern Capitalism*, London: Sage.

Dowler, E. and Turner, S. (2001) *Poverty Bites: food, health and poor families*, London: CPAG.

Doyal, L. (1979) *The Political Economy of Health*, Second Edition, 1994, London: Pluto.

Duménil, G. and Levy, D. (1993) *The Economics of the Profit Rate*, London: Edward Elgar.

Duménil, G. and Levy, D. (2001a) 'The nature and contradictions of neoliberalism', in L. Panitch and C. Leys (eds) *Socialist Register 2002*, London: Merlin.

Duménil, G. and Levy, D. (2001b) 'Costs and benefits of neoliberalism: a class analysis', *Review of International Political Economy*, 8 (4): 578–607.

Duncan, S. (1994) 'Theorising differences in patriarchy', *Environment and Planning A*, 26: 1177–95.

Duncan, S. and Pfau-Effinger, B. (eds) (2000) *Gender, Economy and Culture in the European Union*, London: Routledge.

Dunford, M. (1994) 'Winners and losers: the new map of economic inequality in the European Union', *European Urban and Regional Studies*, 1 (2): 95–114.

Dunford, M. and Perrons, D. (1983) *The Arena of Capital*, London: Macmillan.

Dunnigan, M. and Pollock, A. (2003) 'Downsizing of acute in-patient beds associated with the private finance initiative: Scotland's case study', *British Medical Journal*, 326: 905–8.

Eaden, J. and Renton, D. (2002) *The Communist Party of Great Britain since 1920*, Basingstoke: Palgrave Macmillan.

Ehrenreich, B. and English, D. (1989) *For Her Own Good*, London: Anchor.

Ehrenreich, B. and Hochschild, A. (eds) (2003) *Global Woman: nannies, maids and sex workers in the new economy*, London: Granta.

Eisenschitz, A. and Gough, J. (1993) *The Politics of Local Economic Policy*, Basingstoke: Macmillan.

Eisenschitz, A. and Gough, J. (1996) 'The construction of mainstream local economic initiatives: mobility, socialisation and class relations', *Economic Geography*, 76 (2): 178–95.

Eisenschitz, A. and Gough, J. (1997) 'The division of labour, capitalism and socialism: an alternative to Sayer', *International Journal of Urban and Regional Research*, 21 (1): 23–37.

Esping-Andersen, G. (1990) *Three Worlds of Welfare*, Cambridge: Polity Press.

Etherington, D. and Jones, M. (2004) 'Welfare-through-work and the re-regulation of labour markets in Denmark', *Capital and Class*, 83: 19–45.

Etzioni, A. (1993) *The Spirit of Community,* New York: Crown.

Fagan, R. (1991) 'Industrial policy and the macroeconomic environment', *Australian Geographer,* 22: 102–5.

Fainstein, N. (1993) 'Race, class and segregation: discourses about African Americans', *International Journal of Urban and Regional Research,* 17 (3): 384–403.

Fainstein, S. (1994) *The City Builders,* Oxford: Blackwell.

Fainstein, S., Gordon, I. and Harloe, M. (1992) *Divided Cities,* Oxford: Blackwell.

Fairlie, S. (1996) *Low Impact Development,* Charlbury: Jon Carpenter.

Faulks, K. (1998) *Citizenship in Modern Britain,* Edinburgh: Edinburgh University Press.

Ferguson, I. (2000) 'Identity politics or class struggle? The case of the mental health users' movement', in M. Lavalette and G. Mooney (eds) *Class Struggle and Social Welfare,* London: Routledge.

Fine, B. and Harris, L. (1985) *Peculiarities of the British Economy,* London: Lawrence and Wishart.

Finn, D. (1987) *Training without Jobs,* Basingstoke: Macmillan.

Fishman, R. (1987) *Bourgeois Utopias,* New York: Basic Books.

Fitzpatrick, P. (2003) 'Poor, excluded and forgotten: asylum seekers and the welfare state', *Poverty,* 115: 12–16.

Flaherty, J.,Veit-Wilson, J. and Dornan, P. (2004) *Poverty: the facts,* 5th edn, London: Child Poverty Action Group.

Flecker, J. (2002) 'The European Right and working life: from ordinary miseries to political disasters', in L. Panitch and C. Leys (eds) *Socialist Register 2003,* London: Merlin.

Fletcher, B. and Gapasin, F. (2002) 'The politics of labour and race in the USA', in L. Panitch and C. Leys (eds) *Socialist Register 2003,* London: Merlin.

Foley, M. and Edwards, B. (1998) 'Civil society and social capital: beyond Putnam', *American Behavioural Scientist,* 42 (2): 124–39.

Folwell, K. (1999) *Getting the Measure of Social Exclusion,* London: London Research Centre.

Forrest, R. and Kearns, A. (1999) *Joined-up Places? Social cohesion and neighbourhood regeneration,* York: Joseph Rowntree Foundation.

Forrest, R. and Kearns, A. (2001) 'Social cohesion, social capital and the neighbourhood', *Urban Studies,* 38 (12): 2125–43.

Fraser, N. (1995) 'From redistribution to recognition: dilemmas of justice in a "post-socialist" age', *New Left Review,* 212: 68–93.

Frege, C. and Kelly, J. (2003) 'Union revitalization strategies in comparative perspective', *European Journal of Industrial Relations,* 9 (1): 7–24.

Freire, P. (1972) *Pedagogy of the Oppressed,* Harmondsworth: Penguin.

Friedman, A. (1977) *Industry and Labour,* Basingstoke: Macmillan.

Fung, A. (2003) 'Deliberative democracy, Chicago style; grass-roots governance in policing and public education', in A. Fung and E. Wright (eds) *Deepening Democracy,* London: Verso.

Fung, A. and Wright, E. (2003) 'Thinking about empowered participatory governance', in A. Fung and E. Wright (eds) *Deepening Democracy,* London: Verso.

Gallie, D. and Vogler, C. (1990) 'Unemployment and attitudes to work', *Working Paper No. 18, Social Change and Economic Life Initiative,* Oxford: Nuffield College.

Gallin, D. (2001) 'Propositions on trade unions and informal employment in times of globalisation', in P. Waterman and J. Wills (eds) *Place, Space and New Labour Internationalisms,* Oxford: Blackwell.

Garrahan, P. and Stewart, P. (1992) *The Nissan Enigma,* London: Mansell.

Geras, N. (1998) *The Contract of Mutual Indifference*, London: Verso.

Giddens, A. (1991) *Modernity and Self-identity*, Cambridge: Polity Press.

Giddens, A. (1998) *The Third Way,* Cambridge: Polity Press.

Gill, S. (2002) *Power and Resistance in the New World Order*, Basingstoke: Palgrave Macmillan.

Ginn, J. and Arber, S. (1996) 'Patterns of employment, gender and pensions: the effect of work history on older women's non-state pensions', *Work, Employment and Society*, 10: 469–90.

Girouox, H. (2004) *The Terror of Neoliberalism*, Boulder, CO: Paradigm Publishers.

Glass, N. (2005) 'Surely some mistake?', *The Guardian*, 5 January.

Glendinning, C. and Millar, J. (1992) *Women and Poverty in Britain in the 1990s,* Hemel Hempstead: Harvester Wheatsheaf.

Glennerster, H. (2000) *US Poverty Studies and Poverty Measurement,* London: Centre for Analysis of Social Exclusion.

Glyn, A. (1988) 'The economic case against pit closures', in D. Cooper and T. Hopper (eds) *Debating Coal Closures,* Cambridge: Cambridge University Press.

Goheen, P. (1998) 'Public space and the geography of the modern city', *Progress in Human Geography*, 22 (4): 479–96.

Golding, P. and Middleton, S. (1982) *Images of Welfare: press and public attitudes to welfare*, Oxford: Blackwell.

Goode, J., Callender, C. and Lister, R. (1998) *Purse or Wallet?* York: Joseph Rowntree Foundation/Policy Studies Institute.

Goodwin, M. (2003) 'Partnerships and the Third Way: "the best possible political shell"'? Paper given at the Institute of British Geographers/RSG Conference, London, September 2003.

Gordon, D. (2000) 'Measuring absolute and overall poverty', in D. Gordon and P. Townsend (eds) *Breadline Europe*, Bristol: Polity Press.

Gordon, D. and Pantazis, C. (1997) *Breadline Britain in the 1990s*, Aldershot: Ashgate.

Gordon, D. and Townsend, P. (2000) *Breadline Europe,* Bristol: Policy Press.

Gordon D., Townsend, P., Levitas, R., Pantazis, C., Payne, S. and Patsios D. (2000) *Poverty and Social Exclusion in Britain*, York: Joseph Rowntree Foundation.

Gordon, M. (2002) 'The contribution of the community cooperatives of the Highlands and Islands of Scotland to the Development of the social economy', *Journal of Rural Cooperation*, 30 (2): 95–117.

Gorz, A. (1982) *Farewell to the Working Class*, London: Pluto.

Gosling, A., Machin, S. and Meghir, C. (1994) *The Changing Distribution of Wages in the UK 1966–1992,* London: Institute of Fiscal Studies.

Gottdiener, M. (1987) *The Decline of Urban Politics,* Beverly Hills, CA: Sage.

Gough, I. (1982) *The Political Economy of the Welfare State*, London: Macmillan.

Gough, J. (1986) 'Industrial policy and socialist strategy: restructuring the unity of the working class', *Capital and Class*, 29: 58–82.

Gough, J. (1991) 'Structure, system and contradiction in the capitalist space economy', *Environment and Planning D: Society and Space*, 9: 433–91.

Gough, J. (1992) 'Workers' competition, class relations and space', *Environment and Planning D: Society and Space*, 10: 265–86.

Gough, J. (1996a) 'Not flexible accumulation: contradictions of value in contemporary economic geography. Part 1: workplace and inter-firm relations', *Environment and Planning A*, 28: 2063–79.

Gough, J. (1996b) 'Not flexible accumulation: contradictions of value in contemporary economic geography. Part 2: regional regimes, national regulation and political strategy', *Environment and Planning A*, 28: 2179–200.

Gough, J. (2002) 'Neoliberalism and socialisation in the contemporary city: opposites, complements and instabilities', *Antipode*, 34 (3): 405–26.

Gough, J. (2003a) *Work, Locality and the Rhythms of Capital*, London: Continuum/Routledge.

Gough, J. (2003b) 'The genesis and tensions of the English regional development agencies: class relations and scale', *European Urban and Regional Studies*, 10 (1): 23–38.

Gough, J. (2004a) 'Changing scale as changing class relations: variety and contradiction in the politics of scale', *Political Geography*, 23 (2): 185–211.

Gough, J. (2004b) 'National regimes of the more-developed countries and the generation of poverty and exclusion', Working Paper, Division of Geography, Northumbria University. Online. Available at http://www.sheffield.ac.uk/tmp/about/staff/jamie_gough

Gough, J. and Eisenschitz, A. (1996a) 'The contradictions of neo-Keynesian local economic strategies', *Review of International Political Economy*, 3 (3): 434–58.

Gough, J. and Eisenschitz, A. (1996b) 'The modernisation of Britain and local economic policy: promise and contradictions', *Environment and Planning D: Society and Space*, 14: 203–19.

Gough, J. and Eisenschitz, A. (1998) 'Theorising the state in local economic governance', *Regional Studies*, 32 (8): 759–68.

Gough, J. and Macnair, M. (1985) *Gay Liberation in the Eighties*, London: Pluto.

Gowan, P. (1999) *The Global Gamble*, London: Verso.

Greater London Authority (GLA) (2002) *London Divided*, London: GLA.

Greater London Council (GLC) (1985) *London Industrial Strategy*, London: GLC.

Greed, C. (1994) *Women and Planning*, London: Routledge.

Green, A. (1994) *The Geography of Poverty and Wealth*, Institute for Employment Research, Warwick: University of Warwick.

Green, A. and Owen, D. (1998) *Where are the Jobless?*, Bristol: Policy Press.

Green, D. (1998) *Benefit Dependency: how welfare undermines independence*, London: Institute of Economic Affairs.

Green, F. (ed.) (1989) *The Restructuring of the UK Economy*, Hemel Hempstead: Harvester Wheatsheaf.

Greenberg, D. (1988) *The Construction of Homosexuality*, Chicago, IL: University of Chicago Press.

Grieve Smith, J. (2000) *Closing the Casino: reform of the global financial system*, London: Fabian Society.

Guérin, D. (1973) *Fascism and Big Business*, New York: Monad Press.

Gutman, H. (1976) *Work, Culture and Society in Industrialising America*, New York: Knopf.

Habermas, J. (1976) *Legitimation Crisis*, London: Heinemann.

Hall, P., Thomas, R., Gracey, H. and Drewett, R. (1973) *The Containment of Urban England*, London: Allen and Unwin.

Hall, S. (2003) 'The "Third Way" revisited: "New" Labour, spatial policy and the National Strategy for Neighbourhood Renewal', *Planning, Practice and Research*, 18 (4): 265–77.

Hall, S. and Hickman, P. (2002) 'Neighbourhood renewal and urban policy: a comparison of new approaches in England and France', *Regional Studies*, 36 (6): 691–6.

Hall, S. and Jacques, M. (eds) (1991) *New Times,* London: Verso.

Hall, T. and Hubbard, P. (1996) 'The Entrepreneurial City: new urban politics, new urban geographies', *Progress in Human Geography*, 20 (2): 153–74.

Hambledon, R. (2003) 'The new city management', in R. Hambledon, H. Savitch and M. Stewart (eds) *Globalism and Local Democracy*, Basingstoke: Palgrave.

Hamnett, C. (1998) *Winners and Losers: the homeownership market in modern Britain,* London: University College London Press.

Hamnett, C. (2002) *Unequal City,* London: Routledge.

Handler, J. and Hasenfeld, Y. (1991) *The Moral Construction of Poverty: welfare reform in America*, Newbury Park, CA: Sage.

Handler, J. and Hasenfeld, Y. (1997) *We the Poor People: work, poverty and welfare*, New Haven, CT: Yale University Press.

Hannington, W. (1937) *The Problem of the Distressed Areas*, London: Gollancz.

Haq, J. (2004) Personal communication.

Harris, N. (1986) 'What to do with London? the strategies of the GLC in 1981–6', *International Socialism*, 2: 113–34.

Harris, N. (1995) *The New Untouchables: immigration and the new world worker,* Harmondsworth: Penguin.

Harrison, B. (1997) *Lean and Mean,* New York: Guilford Press.

Harvey, D. (1973) *Social Justice and the City,* London: Edward Arnold.

Harvey, D. (1982) *The Limits to Capital*, Oxford: Blackwell.

Harvey, D. (1989) *The Urban Experience*, Oxford: Blackwell.

Harvey, D. (1992) 'Social justice, postmodernism and the city', *International Journal of Urban and Regional Studies*, 16 (4): 588–601.

Harvey, D. (2000) *Spaces of Hope,* Berkeley, CA: University of California Press.

Harvey, D. (2003) *The New Imperialism*, Oxford: Oxford University Press.

Hayden, D. (1980) 'What would a non-sexist city be like?', in C. Stimpson, E. Dixler, M. Nelson and K. Yatrakis (eds) *Women and the American City*, Chicago, IL: University of Chicago Press.

Hayter, T. (1997) *Urban Politics: accommodation or resistance?*, Nottingham: Spokesman.

Hayter, T. (2004) *Open Borders: the case against immigration controls,* London: Pluto.

Hayton, K. (2000), 'Scottish Community Business: an idea that has had its day?', *Policy and Politics*, 28 (2): 193–206.

Healey, P. (1998), 'Building institutional capacity through collaborative approaches to urban planning', *Environment and Planning A*, 30: 1531–46.

Hebbert, M. (1980) *The Inner City Problem in Historical Context,* London: Social Science Research Council.

Hennessy, R. and Ingraham, C. (eds) (1997) *Materialist Feminism*, New York: Routledge.

Hensman, R. (1991) *A Common Cause: a discussion of common interests of victims of corporate irresponsibility: Bhopal and Britain*, London: Workers' Education Assocation.

Hewitt, M. (2002) 'New Labour and the redefinition of social security', in M. Powell (ed.) *Evaluating New Labour's Welfare Reforms*, Bristol: Policy Press.

Hey, V. (1986) *Patriarchy and Pub Culture*, London: Tavistock.

Hills, J., Le Grand, J. and Piachaud, D. (eds) (2002) *Understanding Social Exclusion,* Oxford: Oxford University Press.

Hirst, P. and Zeitlin, J. (1989) *Reversing Industrial Decline?*, Oxford: Berg.

Hodgson, G. (1984) *The Democratic Economy*, Harmondsworth: Penguin.

Hodgson, G. (1998) *Economics and Utopia: why the learning economy is not the end of history*, London: Routledge.

Hoggart, R. (1957) *The Uses of Literacy: aspects of working class life*, London: Chatto and Windus.

Howarth, C., Kenway, P., Palmer, G. and Morelli, R. (1999) *Monitoring Poverty and Social Exclusion 1999*, York: Joseph Rowntree Foundation.

Howitt, R. (1993) '"A world in a grain of sand": towards a reconceptualisation of geographical scale', *Australian Geographer*, 24: 33–44.

Hudson, R. and Williams, A. (1995) *Divided Britain*, 2nd edn, Chichester: Wiley.

Hudson, W. (1988) 'Labour market changes and new forms of work in "old" industrial regions', in D. Massey and J. Allen (eds) *Uneven Redevelopment*, London: Hodder and Stoughton.

Hughes, R. (1988) *The Fatal Shore*, London: Pan.

Humphrey, J. (2003) 'New Labour and the regulatory reform of social care', *Critical Social Policy*, 23 (1): 5–24.

Husan, R. (2003) 'Critical remarks on cultural aspects of Asian ghettos in modern Britain', *Capital and Class*, 81: 103–34.

Hutton, W. (1995) 'The 30–30–40 society', *Regional Studies*, 29 (8): 719–21.

Illich, I. (1983) *Disabling Professions*, London: Marion Boyars.

Ingham, G. (1984) *Capitalism Divided? City and industry in British social development*, Basingstoke: Macmillan.

Itoh, M. (1995) *Political Economy for Socialism*, Basingstoke: Macmillan.

Jackson, P. (1994) 'Influences on commitment to employment and commitment to work', in A. Bryson and S. McKay (eds) *Is It Worth Working?*, London: Policy Studies Institute.

Jessop, B. (2000) 'The state and the contradictions of the knowledge-driven economy', in J. Bryson, P. Daniels, N. Henry and J. Pollard (eds), *Knowledge, Space, Economy*, London: Routledge.

Johnston, L., MacDonald, R., Mason, P., Ridley, R. and Webster, C. (2000), *Snakes and Ladders: young people, transitions and social exclusion*, York: Policy Press.

Jones, C. and Novak, T. (1999) *Poverty, Welfare and the Disciplinary State*, London: Routledge.

Jones, M. and Gray, A. (2001), 'Social capital or local workfares? Reflections on Employment Zones', *Local Economy*, 16 (3): 178–86.

Judd, D. (1995) 'The rise of the new walled cities', in D. Liggett and D. Perry (eds) *Spatial Practices*, Thousand Oaks, CA: Sage.

Judd, D. (1999) 'Constructing the tourist bubble', in D. Judd and S. Fainstein (eds) *The Tourist City*, New Haven, CT: Yale University Press.

Kakios, M. and van der Velden, J. (1984) 'Migrant communities and class politics: the Greek community in Australia', in G. Bottomley and M. de Lepervanche (eds) *Ethnicity, Class and Gender in Australia*, Sydney: Allen and Unwin.

Katz, I. (2001) *The Long Weekend: combating unemployment during the inter-war years*, London: Hetherington Press.

Kearns, G. and Philo, C. (1993) *Selling Places*, Oxford: Pergamon.

Kempson, E. and Whyley, C. (1999) *Kept Out or Opted Out? Understanding and combating financial exclusion*, Bristol: Policy Press.

Kenney, M. (1997) 'Value creation in the late twentieth century: the rise of the knowledge worker', in J. Davis, T. Hirschl and M. Stack (eds) *Cutting Edge: technology, information, capitalism and social revolution*, London: Verso.

Kerbo, H. (1983) *Social Stratification and Inequality: class conflict in the United States*, New York: McGraw Hill.

Kincaid, J., Samuel, R. and Slater, E. (1962) 'But nothing happens', *New Left Review*, 13–14: 1–13.

Klinenberg, E. (1999) 'Denaturalising disaster: a social autopsy of the 1995 Chicago heat wave', *Theory and Society*, 28: 239–95.

Knopp, L. (1990) 'Some theoretical implications of gay involvement in the urban land market', *Political Geography Quarterly*, 9: 337–52.

Knox, P. (1995) *Urban Social Geography*, 3rd edn, Harlow: Longman.

Kotz, D., McDonough, T. and Reich, M. (eds) (1994) *Social Structures of Accumulation*, Cambridge: Cambridge University Press.

Kreitman, N., Carstairs, V. and Duffy, J. (1991), 'Association of age and social class with suicide among men in Great Britain', *Journal of Epidemiology and Public Health*, 45: 195–202.

Krumholz, N. (1996) 'A retrospective view of equity planning: Cleveland, 1969–1979', in S. Campbell and S. Fainstein (eds) *Readings in Planning Theory*, Oxford: Blackwell.

Kuhn, A. and Wolpe, A-M. (1978) *Feminism and Materialism*, London: Routledge.

Kumar, S. (1997) *Accountability in The Contract State*, York: Joseph Rowntree Foundation.

Laing, W. (2004), *Calculating a Fair Price for Care*, 2nd edn, York: Policy Press.

Langley, P. (2004) 'Retirement, responsibility and risk: the making of investor subjects in Anglo-American pensions', Paper presented at the British International Studies Association Annual Conference, University of Warwick, December 2004.

Lash, S. and Urry, J. (1987) *The End of Organised Capitalism*, Cambridge: Polity Press.

Lash, S. and Urry, J. (1994) *Economies of Signs and Spaces*, London: Sage.

Lavalette, M. and Mooney, G. (2000) '"No poll tax here!": the Tories, social policy and the great poll tax rebellion, 1987–1991, in M. Lavalette and G. Mooney (eds) *Class Struggle and Social Welfare*, London: Routledge.

Lawrence, F. (2005) 'The Third Way's dirtiest secret', *The Guardian*, 3 February.

Layard, R. (1997) 'Preventing long-term unemployment', in J. Philpott (ed.) *Working for Full Employment*, London: Routledge.

Leadbeater, C. (1997) *The Rise of the Social Entrepreneur*, London: Demos.

Lee, R. (1996) 'Moral money? LETS and the social construction of local economic geographies in Southeast England', *Environment and Planning A*, 28: 1377–94.

Leira, A. (2002) *Working Parents and the Welfare State*, Cambridge: Cambridge University Press.

Lembcke, J. (1993) 'Class formation and class capacities: a new approach to the study of labor and the labor process', in B. Berberoglu (ed.) *The Labor Process and Control of Labor*, Westport, CT: Praeger.

Levitas, R. (1986) 'Competition and compliance: the utopias of the New Right', in R. Levitas (ed.) *The Ideology of the New Right*, Cambridge: Polity Press.

Levitas, R. (1998) *The Inclusive society? Social exclusion and New Labour*, Basingstoke: Palgrave Macmillan.

Levitas, R. (2001) 'Unequal Britain', *Renewal*, 9 (4): 15–24.

Lewis, O. (1979) *The Children of Sanchez*, London: Vintage Books.

Leyshon, A. and Thrift, N. (1996) 'Financial exclusion and the shifting boundaries of the financial system', *Environment and Planning A*, 28: 1150–6.

Leyshon, A., Thrift, N. and Pratt, J. (1998) 'Reading financial services: texts, consumers and financial literacy', *Environment and Planning D*, 16: 29–55.

Li, Y., Savage, M. and Pickles, A. (2003) 'Social capital and social exclusion in England and Wales', *British Journal of Sociology*, 54 (4): 497–562.

Lindsay, C. (2005) 'Employment and unemployment estimates for 1971 to 1991', *Labour Market Trends*, London: Office of National Statistics.

Lipietz, A. (1992) *Towards a New Economic Order: post-Fordism, democracy and ecology*, London: Pluto Press.

Little, J, Peake, L. and Richardson, P. (1988) *Women in Cities*, London: Macmillan.

Lowe, S., Walker, R. and Hughes, D. (eds) (1991) *A New Century of Social Housing*, London: Continuum.

Lupton, R. and Power, A. (2002) 'Social exclusion and neighbourhoods', in J. Hills, J. Le Grand and D. Piachaud (eds) *Understanding Social Exclusion*, Oxford: Oxford University Press.

Lyon, D. (2001) *Surveillance Society*, Milton Keynes: Open University Press.

McCahill, M. (2002) *The Surveillance Web: the rise of CCTV in an English city*, Cullompton: Willan Publishing.

McCahill, M. and Norris, C. (2003) 'Victims of surveillance', in P. Davies, P. Francis and V. Jupp (eds) *Victimisation: theory, research and policy*, Basingstoke: Palgrave.

McCulloch, A. (2004) 'Localism and its neoliberal application: a case study of West Gate New Deal for Communities in Newcastle upon Tyne, UK', *Capital and Class*, 83: 133–65.

McDowell, L. and Massey, D. (1984) 'A woman's place?' in D. Massey and J. Allen (eds) *Geography Matters!*, Cambridge: Cambridge University Press.

McGregor, A., Ferguson, Z., Fitzpatrick, I., McConnachie, M. and Richmond, K. (1997) *Bridging the Jobs Gap*, York: York Publishing.

Mcilwaine, C. (1998) 'Civil society and development geography', *Progress in Human Geography*, 22 (3): 415–24.

Macintyre, S. (1980) *Little Moscows: communism and working-class militancy in inter-war Britain*, London: Croom Helm.

McKay, A. and van Every, J. (2000) 'Gender, family and income maintenance: a feminist case for citizens' basic income', *Social Politics*, 7 (2): 266–84.

McKay, D. (1976) *Housing and Race in Industrial Society*, New York: Rowan and Littlefield.

McKay, S. (2003) 'Reforming pensions: investing in the future', in J. Millar (eds) *Understanding Social Security*, Bristol: Polity Press.

McKibbin, R. (1998) *Classes and Cultures: England 1918–1951*, Oxford: Oxford University Press.

Mackintosh, M. (1987) 'Planning the public sector: an argument from the case of transport in London', in A. Cochrane (ed.) *Developing Local Economic Strategies*, Milton Keynes: Open University Press.

Mackintosh, M. and Wainwright, H. (1987) *A Taste of Power*, London: Verso.

McLaughlin, E., Millar, J. and Cooke, K. (1989) *Work and Welfare Benefits*, Aldershot: Gower.

MacLeod, G. (2002) 'From urban entrepreneurialism to a "revanchist city"? On the spatial injustices of Glasgow's renaissance', *Antipode*, 34 (3): 602–24.

Madanipour, A. (1998) 'Social exclusion and space', in G. Cars, A. Madanipour and J. Allen (eds) *Social Exclusion in European Cities*, London: Jessica Kingsley.

Mandel, E. (1969) *The Marxist Theory of the State*, New York: Pathfinder Press.

Mandel, E. (1978) *Late Capitalism*, London: Verso.

Mandel, E. (1986) *The Meaning of the Second World War*, London: Verso.

Mann, K. (1992) *The Making of an English Underclass?*, Milton Keynes: Open University Press.

Marcus, S. (1969) *The Other Victorians: a study of sexuality and pornography in mid-nineteenth century England*, London: Corgi.

Marcuse, H. (1964) *One Dimensional Man*, London: Routledge and Kegan Paul.

Markusen, A. (1987) *Regions: the economics and politics of territory*, Totowa, NJ: Rowan and Littlefield.

Marquand, D. (1999) *The Progressive Dilemma: from Lloyd George to Blair*, London: Phoenix.

Martin, R., Nativel, C. and Sunley, P. (2003) 'The local impact of the New Deal: does geography make a difference?', in R. Martin and P. Morrison (eds) *Geographies of Labour Market Inequality*, London: Routledge.

Marx, K. (1970) *Capital, Volume 1*, London: Lawrence and Wishart.

Massey, D. (1974) *Towards a Critique of Industrial Location Theory*, London: Centre for Environmental Studies.

Massey, D. (1979) 'In what sense a regional problem?', *Regional Studies*, 3 (2): 233–44.

Massey, D. (1984) *Spatial Divisions of Labour*, Basingstoke: Macmillan.

Mayes, D., Berghman, J. and Salais, R. (2001) *Social Exclusion and European Policy*, 2nd edn, Cheltenham: Edward Elgar.

Mayo, M. (2000) *Cultures, Communities, Identities*, Basingstoke: Palgrave.

Mead, L. (1997a) *The New Paternalism: supervisory approaches to poverty*, Washington, DC: Brookings Institution Press.

Mead, L. (1997b) 'From welfare to work: lessons from America', in A. Deacon (ed.) *From Welfare to Work*, London: Institute of Economic Affairs.

Medoff, P. and Sklar, H. (1994) *Streets of Hope*, Boston, MA: South End Press.

Mee, K. (1994) 'Dressing up the suburbs: representations of Western Sydney', in K. Gibson and S. Watson (eds) *Metropolis Now*, London: Pluto.

Mee, K. and Dowling, R. (2000) 'Tales of the city: Western Sydney at the end of the millenium', in J. Connell (ed.) *Sydney*, South Melbourne: Oxford University Press.

Meiskins Wood, E. (1995) *Democracy against Capitalism*, Cambridge: Cambridge University Press.

Meiskins Wood, E. (2002) *The Origins of Capitalism: a longer view*, London: Verso.

Mellor, M., Stirling, J. and Hannah, J. (1998) *Worker Cooperatives*, Milton Keynes: Open University Press.

Melossi, D. and Pavarini, M. (1981) *The Prison and the Factory*, London: Macmillan.

Metzaros, I. (1970) *Marx's Theory of Alienation*, London: Merlin.

Miles, D. and Timmerman, A. (1999) 'Risk sharing and transition costs in the reform of pension systems in Europe', *Economic Policy*, 29: 253–86.

Miliband, R. (1961) *Parliamentary Socialism*, London: Allen and Unwin.

Millar, F. (2004) 'The links between housing and educational outcomes cannot be ignored any longer', *The Guardian*, 18 January.

Millar, J. (2003a) 'Social security: means and ends', in J. Millar (ed.) *Understanding Social Security*, Bristol: Polity Press.

Millar, J. (2003b) 'From wage replacement to wage supplement: benefits and tax credits', in J. Millar (ed.) *Understanding Social Security*, Bristol: Polity Press.

Millar, J. and Gardiner, K. (2004) *Low Pay, Household Resources and Poverty*, York: Joseph Rowntree Foundation.

Miller, S., Rein, M. and Levitt, P. (1995) 'Community action in the United States', in G. Craig and M. Mayo (eds) *Community Empowerment*, London: Zed Books.

Mingione, E. (1988) 'Work and informal activities in urban south Italy', in R. Pahl (ed.) *On Work,* Oxford: Blackwell.

Mingione, E. (1996) 'Urban poverty in the advanced industrial world: concepts, analysis and debates', in E. Mingione (ed.) *Urban Poverty and the Underclass: a reader*, Oxford: Blackwell.

Minns, R. (2001) *The Cold War in Welfare: stock markets versus pensions*, London: Verso.

Mitchell, D. (1991) *Income Transfers in Ten Welfare States*, Aldershot: Ashgate.

Mitchell, D. (2003) *The Right to the City: social justice and the fight for public space*, New York: Guilford Press.

Modood, T. (1997) *Ethnic Minorities in Britain: diversity and disadvantage: the fourth national survey of ethnic minorities*, London: Policy Studies Institute.

Mohan, J. (1999) *A United Kingdom?,* London: Arnold.

Monbiot, G. (2000) *Captive State: the corporate takeover of Britain*, Basingstoke: Palgrave Macmillan.

Moody, G. (2001) *Rebel Code*, London: Allen Lane.

Moody, K. (1997) *Workers in a Lean World,* London: Verso.

Morris, L. (1990) *The Workings of the Household,* Cambridge: Polity Press.

Morris-Suzuki, T. (1997) 'Capitalism in the computer age', in J. Davis, T. Hirschl and M. Stack (eds) *Cutting Edge: technology, information, capitalism and social revolution*, London: Verso.

Moses, D. (ed.) (2004) *Genocide and Settler Society*, New York: Berghahn Books.

Moulaert, F. and Cavola, L. (2004) 'Analysing social exclusion in metropolitan areas: combining "general" and "specific" perspectives', Department of Planning, Newcastle University.

Moulaert, F. and Gonzalez, S. (2003) 'Is there life beyond the knowledge-based society? Towards a decelerated approach to learning, creativity and enjoyment in Europe', Department of Planning, Newcastle University.

Moulaert, F. and Nussbaumer, J. (2005) 'The social region: beyond the territorial dynamics of the learning economy', *European Urban and Regional Studies*, 12 (1): 45–64.

Mulgan, G. (1991) *Communication and Control,* Oxford: Polity Press.

Mullan, P. (2002) *The Imaginary Timebomb: why an ageing population is not a social problem*, New York: I.B. Tauris.

Murray, C. (1990) *The Emerging British Underclass*, London: Institute of Economic Affairs.

Nadonovsky, P. (1993) 'Biting inequalities: dental health, the rich and the poor', *The New Review*, 6–12.

Nairn, T. (1972), 'The English working class', in R. Blackburn (ed.), *Ideology in Social Science*, London: Fontana.

National Statistics (2004) 'British Household Panel survey'. Available <http://www.dwp.gov.uk/asd/hbal.asp> (accessed 2 January 2005)

Neocleous, M. (2002) *The Fabrication of Social Order: a critical theory of police power*, London: Pluto.

Nichols, T. (2001) 'The condition of labour – a retrospect', *Capital and Class*, 75, 185–98.

Nickell, S. (2004) 'Poverty and worklessness in Britain', *The Economic Journal*, 114: 1–25.

Nickell, S., Jones, P. and Quintini, G. (2002) 'A picture of job insecurity facing British men', *Economic Journal*, 112: 1–27.

Noble, D. (1984) *Forces of Production: a social history of industrial automation*, Oxford: Oxford University Press.

Noble, M. and Smith, G. (1996) 'Two nations? Changing patterns of income and wealth in two contrasting areas', in J. Hills (ed.) *New Inequalities*, Cambridge: Cambridge University Press.

Oatley, N. (ed.) (1998) *Cities, Economic Competition and Urban Policy*, London: Paul Chapman.

Oatley, N. (1999) 'Developing the social economy', *Local Economy*, 13 (4): 339–45.

O'Connor, J. (1987) *The Meaning of Crisis*, Oxford: Blackwell.

Offe, C. (1984) *Contradictions of the Welfare State*, London: Hutchinson.

O'Gorman, F. (1995) 'Brazilian community development', in G. Craig and M. Mayo (eds) *Community Empowerment*, London: Zed Books.

Ohmae, K. (1995) *The End of the Nation State*, London: HarperCollins.

Oliver, M. and Barnes, C. (1998) *Disabled People and Social Policy*, London: Longman.

Ollman, B. (1993) *Dialectical Investigations*, New York: Routledge.

O'Neill, P. (1997) 'So what is internationalisation? Lessons from restructuring at Australia's "mother plant"', in M. Taylor, and S. Conti (eds) *Interdependent and Uneven Development*, Aldershot: Ashgate.

Onyx, J. and Benton, P. (1995) 'Empowerment and ageing: toward honoured places for crones and sages', in G. Craig and M. Mayo (eds) *Community Empowerment*, London: Zed Books.

Organisation for Economic Cooperation and Development (OECD) (1999) *The Local Dimension of Welfare to Work: an international survey*, Paris: OECD.

Pacione, M. (1997) 'Urban restructuring in Britain's cities', in M. Pacione (ed.) *Britain's Cities*, London: Routledge.

Page, D. (2000) *Communities in the Balance: the reality of social exclusion on housing estates*, York: Joseph Rowntree Foundation.

Pahl, R. and Wallace, C. (1985) 'Forms of work and privatisation on the Isle of Sheppey', in B. Roberts, R. Finnegan and D. Gallie (eds) *New Approaches to Economic Life*, Manchester: Manchester University Press.

Pain, R. (1995) 'Fear of crime and local contexts: elderly people in north east England', *Northern Economic Review*, 24: 96–111.

Pain, R. (1997) 'Social geographies of women's fear of crime', *Transactions of the Institute of British Geographers*, 22 (2): 231–44.

Painter, J. (2002) 'Governmentality and regional economic strategies, in J. Hillier and E. Rooksby (eds) *Habitus: a sense of place*, Aldershot: Ashgate.

Palmer, G., North, J., Carr, J. and Kenway, P. (2004) *Monitoring Poverty and Social Exclusion 2003*, York: Joseph Rowntree Foundation.

Palmer, J. (1972) Introduction, to R. Goodman, *After the Planners*, Harmondsworth: Penguin.

Pearson, G. (1975) *The Deviant Imagination*, London: Macmillan.

Peck, J. (1993) 'The trouble with the Training and Enterprise Councils: a critique of the TECs initiative', *Policy and Politics*, 21: 289–305.

Peck, J. (1995) 'Moving and shaking: business elites, state localism and urban privatism', *Progress in Human Geography*, 19: 16–46.

Peck, J. (2001) *Workfare States*, New York: Guilford.

Peck, J. and Theodore, N. (2000) 'Work first: workfare and the regulation of contingent labour markets', *Cambridge Journal of Economics*, 24: 119–38.

Peck, J. and Theodore, N. (2001) 'Contingent Chicago: restructuring the spaces of temporary labour', *International Journal of Urban and Regional Research*, 25 (3): 471–97.

Peck, J. and Tickell, A. (1995) 'Business goes local: dissecting the "business agenda" in Manchester', *Inernational Journal of Urban and Regional Research*, 19 (1): 55–78.

Peck, J. and Tickell, A. (2002) 'Neoliberalizing space', *Antipode*, 34 (3): 380–404.

Peel, M. (1993) 'A place made poor: the past and future in Elizabeth', *Arena*, 27–39.

Pepper, D. (1993) *Eco-socialism*, London: Routledge.

Perrons, D. (2000) 'Living with risk: labour market transformation, employment policies and social reproduction in the UK', *Economic and Industrial Democracy*, 21 (3): 283–310.

Petras, J. (2000) 'The Third Way: myth and reality', *Monthly Review*, 51: 19–35.

Philo, C. (1995) *Off the Map: the social geography of poverty*, London: Child Poverty Action Group.

Piachaud, D. and Sutherland, H. (2001) 'Child poverty in Britain and the Labour Government', *Journal of Social Policy*, 30 (1): 95–118.

Pinch, S. (1997) *Worlds of Welfare*, London: Routledge.

Pinch, S. (1999) 'Social polarisation: a comparison of evidence from Britain and the US', in J. Bryson, N. Henry, D. Keeble and R. Martin (eds) *The Economic Geography Reader*, Chichester: Wiley.

Piore, M. and Sabel, C. (1984) *The Second Industrial Divide*, New York: Basic Books.

Piven, R. and Cloward, R. (1972) *Regulating the Poor*, London: Tavistock.

Plougmann, P. (2001) 'Internationalisation and the labour market of the EU', in J. Andersen and P. Jensen (eds) *Changing Labour Markets, Welfare Policies and Citizenship*, Bristol: Policy Press.

Pollard, D. (1988) *Give and Take: the losing partnership in Aboriginal poverty*, Sydney: Hale and Iremonger.

Pollins, H. (1964) 'Transport lines and social divisions', in Centre for Urban Studies (ed.) *London: aspects of change*, London: MacGibbon and Kee.

Pollock, A. (2004) *NHS plc*, London: Verso.

Power, A. (2000) 'Poor areas and social exclusion', in A. Power and W. Wilson (eds) *Social Exclusion and the Future of Cities*, CASE Paper 35, London: Centre for Analysis of Social Exclusion.

Prebble, J. (1969) *The Highland Clearances*, Harmondsworth: Penguin.

Propper, C., Burgess, S. and Abraham, P. (2002) *Competition and Quality: evidence from the NHS internal market 1991–1999*, Bristol: CMPO.

Purdue, D., Razzaque, K., Hambleton, R. and Stewart, M. (2000) *Community Leadership in Area Regeneration*, Bristol: Policy Press/Joseph Rowntree Foundation.

Putnam, R. (2000) *Bowling Alone: the collapse and revival of American community*, New York: Simon and Schuster.

Rahman, M., Palmer, G. and Kenway, P. (2001) *Monitoring Poverty and Social Exclusion 2001*, York: Joseph Rowntree Foundation.

Rainnie, A. (1989) *Industrial Relations in Small Firms*, London: Routledge.

Reade, E. (1987) *British Town and Country Planning*, Milton Keynes: Open University Press.

Rex, J. and Moore, R. (1967) *Race, Community and Conflict*, Oxford: Oxford University Press.

Richardson, L. and Mumford, K. (2002) 'Community, neighbourhood, and social infrastructure', in J. Hills, J. Le Grand and D. Piachaud (eds) *Understanding Social Exclusion*, Oxford: Oxford University Press.

Ridley, N. (1988) *The Local Right*, London: Centre for Policy Studies.

Robertson, S. (2005) 'Changing governance/changing equality? Understanding the politics of public–private partnerships in education in Europe'. Available <http://www.genie-tn.net/paper002.htm> (accessed 11 February 2005).

Robins, K. and Webster, F. (1988) 'Athens without slaves . . . or slaves without Athens?', *Science as Culture*, 3: 7–53.

Roemer, J. (1996) *Equal Shares: making market socialism work*, London: Verso.

Rose, M. (1972) *The Relief of Poverty 1834–1914*, London: Macmillan.

Rosenberg, J. (1998) *Against the Odds*, London: Walterton and Elgin Community Homes.

Ross, E. (1983) 'Survival networks: women's neighbourhood sharing in London before World War 1', *History Workshop*, 15: 4–28.

Rowlingson, K. (2003) '"From cradle to grave': social security over the life cycle', in J. Millar (ed.) *Understanding Social Security*, Bristol: Polity Press.

Rowlingson, K. and McKay, S. (2002) *Lone Parent Families*, Harlow: Pearson Education.

Russell, H. (2001) *Local Strategic Partnerships: lessons from New Commitment to Regeneration*, Bristol: Polity Press, Joseph Rowntree Foundation.

Russell, M. and Malhotra, R. (2001) 'Capitalism and disability', in L. Panitch and C. Leys (eds) *Socialist Register 2002*, London: Merlin.

Sainsbury, R. (2003) 'Understanding social security fraud', in J. Millar (ed.) *Understanding Social Security*, Bristol: Polity Press.

Sassen, S. (1991) *The Global City*, Princeton, NJ: Princeton University Press.

Sassen, S. (1993), 'Economy, ethnicity and global cities', *Social Justice*, 20 (3–4): 32–50.

Saul J. (2002) 'Identifying class, classifying difference', in L. Panitch and C. Leys (eds) *Socialist Register 2003*, London: Merlin.

Sawyers, L. and Tabb, W. (eds) (1984) *Sunbelt/Snowbelt: urban development and regional restructuring*, New York: Oxford University Press.

Sayer, A. (1995) *Radical Political Economy: a critique*, Oxford: Blackwell.

Sayer, A. and Walker, R. (1992) *The New Social Division of Labour*, Oxford: Blackwell.

Schlossman, S. (1974) 'The 'culture of poverty', in Ante-Bellum social throught', *Science and Society*, 38: 150–66.

Schneider, F. (2002) 'Size and measurement of the informal economy in 110 countries around the world', Paper presented to a workshop at the Australian National Tax Centre, Canberra, July 2002.

Scott, A. (1998) *Regions in the World Economy*, Oxford: Oxford University Press.

Scruton, R. (2001) *The Meaning of Conservatism*, Basingstoke: Palgrave Macmillan.

Sen, A. (1983) 'Poor, relatively speaking', *Oxford Economic Papers*, 35 (2): 153–69.

Sennett, R. (1998) *The Corrosion of Character: the personal consequences of work in the new capitalism*, New York: Norton.

Sexty, C. (1990) *Women Losing Out: access to housing in Britain today*, London: Shelter.

Shaw, M., Dorling, D., Gordon, D. and Davey-Smith, G. (1999) *The Widening Gap: health inequalities and policy in Britain*, Bristol: Polity Press.

Shucksmith, M. (2000) *Exclusive Countryside: social inclusion and regeneration in rural areas*, York: Joseph Rowntree Foundation.

Shutt, J. (1984) 'Tory Enterprise Zones and the Labour Movement', *Capital and Class*, 23: 19–44.

Sibley, D. (1995) *Geographies of Exclusion*, London: Routledge.

Simmons, M. (1997) *Landscapes of Poverty: aspects of rural England in the late 1990s*, London: Lemos and Crane.

Sinclair, U. (1965) *The Jungle*, Harmondsworth: Penguin.

Sivanandan, A. (1990) 'Britain's Gulags', in A. Sivanandan (ed.) *Communities of Resistance*, London: Verso.

Sivanandan, A. (2001) 'Poverty is the new black', *Race and Class*, 43 (2): 1–5.

Sklar, H. (1995a) 'Imagine a country', in P. Rothenberg (ed.) *Race, Class and Gender in the United States: an integrated study*, New York: St Martin's Press.

Sklar, H. (1995b) *Chaos or Community?* Boston, MA: South End Press.

Smith, D. (1989) '"Not getting on, just getting by": changing prospects in south Birmingham', in P. Cooke (ed.) *Localities*, London: Unwin Hyman.

Smith, N. (1993) 'Homeless/global: scaling places', in J. Bird, B. Curtis, T. Putnam, G. Robertson and L. Tickner (eds) *Mapping the Futures*, London: Routledge.

Smith, N. (1996) *The New Urban Frontier: gentrification and the revanchist city*, London: Routledge.

Smith, T. (1994) *Lean Production: a capitalist utopia?*, Amsterdam: International Institute for Research and Education.

Social Exclusion Unit (SEU) (2001a) *A New Commitment to Neighbourhood Renewal*, London: SEU.

Social Exclusion Unit (SEU) (2001b) *Preventing Social Exclusion*, London: SEU.

Solow, R. (1998) *Work and Welfare*, Princeton, NJ: Princeton University Press.

Speak, S. and Graham, S. (2000) *Service Not Included: social implications of private sector services restructuring in marginalised neighbourhoods*, Bristol: Policy Press.

Sprawl Busters (2004) Available <http://www.sprawl-busters.com> (accessed 12 January 2005).

Stanton, R. (1996) 'The retreat from social need: competitive bidding and local public investment', *Local Economy*, 11 (3): 194–201.

Stedman Jones, G. (1971) *Outcast London*, Oxford: Oxford University Press.

Stedman Jones, G. (1983) *Languages of Class*, Cambridge: Cambridge University Press.

Stone, I (1997) 'British market-led restructuring: a case study of Wearside', in J. Wheelock and A. Mariussen (eds) *Households, Work and Change*, Amsterdam: Kluwer.

Storey, D., Keasey, K., Watson, R. and Wynarczyk, P. (1987) *The Performance of Small Firms*, London: Croom Helm.

Storper, M. (1997) *The Regional World*, New York: Guilford.

Storper, M. and Scott, A. (eds) (1992) *Pathways to Industrialization and Regional Development*, New York: Routledge.

Storper, M. and Walker, R. (1989) *The Capitalist Imperative*, Oxford: Blackwell.

Strangleman, T. and Roberts, I. (1999) 'Looking through the window of opportunity: the cultural cleansing of workplace identity', *Sociology*, 33 (1): 47–67.

Strobel, P. (1996) 'From poverty to exclusion: a wage earning society or a society of human rights?', *International Social Science Journal*, 148: 173–90.

Sunley, P. (2000), 'Pension exclusion in grey capitalism: mapping the pensions gap in Britain', *Transactions of the Institute of British Geography*, 25: 483–501.

Sutcliffe, B. and Glyn, A. (1999) 'Still underwhelmed: indicators of globalisation and their misinterpretation', *Review of Radical Political Economy*, 31: 111–32.

Sutherland, H., Sefton, T. and Piachaud, D. (2003) *Poverty in Britain: the impact of government policy since 1997*, York: Joseph Rowntree Foundation.

Swyngedouw, E., Moulaert, F. and Rodriguez, A. (2002) 'Neoliberal urbanisation in Europe: large scale urban development projects and the new urban policy', *Antipode*, 34 (3): 542–77.

Szereter, S. (2001) 'Social capital, the economy and education in historical perspective', in S. Baron, J. Field and T. Schuller (eds) *Social Capital: critical perspectives*, Oxford: Oxford University Press.

Tams, E. (2002) 'Creating division: creativity, entrepreneurship and gendered inequality: a Sheffield case study', *City*, 6 (3): 393–402.

Taub, R. (1994) *Community Capitalism: the South Shore Bank's strategy for neighbourhood revitalisation*, Boston, MA: Harvard Business School Press.

Taylor, G. and Mathers, A. (2002) 'The politics of European integration: a European labour movement in the making?', *Capital and Class*, 78: 39–60.

Taylor, I. (1999) *Crime in Context*, Cambridge: Polity Press.

Taylor, I., Evans, K. and Fraser, P. (1996) *A Tale of Two Cities*, London: Routledge.

Taylor-Gooby, P. (1999) 'Policy change at a time of retrenchment: recent pension reform in France, Germany, Italy and the UK', *Social Policy and Administration*, 33 (1): 1–19.

Tebbutt, M. (1983) *Making Ends Meet: pawnbroking and working class credit*, Leicester: Leicester University Press.

Theodore, N. and Peck, J. (2000) 'Searching for best practice in welfare-to-work: the means, the method and the message', *Policy and Politics*, 29 (1): 81–98.

Therborn, G. (1986) *Why Some People are More Unemployed than Others*, London: Verso.

Thomas, B. and Dorling, D. (2004) *Know Your Place: housing wealth and inequality in Great Britain 1980–2003 and beyond*, London: Shelter.

Thompson, E. (1967) 'Time, work discipline and industrial capitalism', *Past and Present*, 38: 56–103 .

Thompson, E. (1968) *The Making of the English Working Class*, Harmondsworth: Penguin.

Tilly, C. (1996) *The Good, the Bad and the Ugly: good and bad jobs in the US at the Millennium*, New York: Russell Sage Foundation.

Townsend, P. (1979) *Poverty in the United Kingdom*, London: Allen Lane.

Tudor Hart, J. (1971) 'The inverse care law', *Lancet*, i: 405–12.

Turner, J. and Fichter, R. (1972) *Freedom to Build*, New York: Macmillan.

Turok, I. and Edge, N. (1999) *The Jobs Gap in Britain's Cities*, Bristol: Policy Press.

Turok, I. and Webster, D. (1998) 'The New Deal: jeopardised by the geography of unemployment?', *Local Economy*, 12 (4): 309–28.

Vail, J., Wheelock, J. and Hill, M. (eds) (1999) *Insecurity Times*, London: Routledge.

Vincent, D. (1991) *Poor Citizens: the state and the poor in twentieth century Britain*, Harlow: Longman.

Wacquant, L. (1996) 'Red belt, black belt: racial division, class inequality and the state in the French urban periphery and the American ghetto', in E. Mingione (ed.) *Urban Poverty and the Underclass: a reader*, Oxford: Blackwell.

Wadhams, C. (2002) *Neighbourhoods: a guide to neighbourhood management for Registered Social Landlords*, London: Harding Housing Association.

Wainwright, H. (1994) *Arguments for a New Left*, Oxford: Blackwell.

Waldinger, R. and Lapp, M. (1999) 'Back to the sweatshop or ahead to the informal sector?', in J. Bryson, N. Henry, D. Keeble and R. Martin (eds) *The Economic Geography Reader*, Chichester: Wiley.

Walker, C. (2000) 'The forms of privatisation of social security in Britain', in J. Dixon and M. Hyde (eds) *The Marketization of Social Security*, Westport, CT: Forum Books.

Walker, R. (1978) 'Two sources of uneven development under advanced capitalism:

spatial differentiation and capital mobility', *Review of Radical Political Economy*, 10: 28–37.

Walker, R. (1981) 'A theory of suburbanization: capitalism and the construction of urban space in the United States', in M. Dear and A. Scott (eds) *Urbanisation and Urban Planning in Capitalist Society*, London: Methuen.

Walklate, S. and Evans, K. (1999) 'You're alright if you come from round here', in P. Hoggett (ed.) *Contested Communities*, Bristol: Policy Press.

Ward, C. (1976) Preface to J. Turner, *Housing by People*, London: Marion Boyars.

Waterman, P. (1998) *Globalization, Social Movements and the New Internationalisms*, London: Mansell.

Webster, D. (2002) 'Unemployment: how official statistics distort analysis and policy, and why', *Radical Statistics*, 079/ 080: 96–127.

Weiss, L. (1988) *Creating Capitalism: the state and small businenss since 1945*, Oxford: Blackwell.

Weiss, L. and Hobson, J. (1995) *States and Economic Development*, Cambridge: Polity Press.

Went, R. (2000) *Globalization: neoliberal challenges, radical responses*, London: Pluto.

Western, B. and Beckett, K. (1999) 'How unregulated is the US labor market? The penal system as a labor market institution', *American Journal Sociology*, 104: 1030–60.

White, M. and Forth, J. (1998) *Pathways through Unemployment*, York: York Publishing Services.

Whitehead, M. (1992) *Inequalities in Health*, Harmondsworth: Penguin.

Widgery, D. (1993) *Some Lives: a GP's East End*, London: Simon and Schuster.

Wiewel, W. and Gills, D. (1995) 'Community development, organizational capacity and US urban policy: lessons from the Chicago experience 1983–93', in G. Craig and M. Mayo (eds) *Community Empowerment*, London: Zed Books.

Wilkinson, M. and Craig, G. (2002) *New Roles for Old: local authority members and partnership working*, York: York Publishing.

Williams, C. and Windebank, J. (1998) *Informal Employment in the Advanced Economies*, London: Routledge.

Williams, C. and Windebank, J. (2002) *Poverty and the Third Way*, London: Routledge.

Williams, R. (1975) *The Country and the City*, Oxford: Oxford University Press.

Willis, P. (1977) *Learning to Labour: how working-class kids get working-class jobs*, London: Saxon House.

Wills, J. (1998) 'Taking on the cosmo cops? Experiments in transnational labour organisation', *Economic Geography*, 74 (2): 111–30.

Wills, J. and Simms, M. (2004) 'Building reciprocal community unionism in the UK', *Capital and Class*, 82: 59–84.

Wilson, E. (1977) *Women and the Welfare State*, London: Tavistock.

Wilson, E. (1980) *Only Halfway to Paradise: women in postwar Britain 1945–1968*, London: Tavistock.

Work Foundation (2004) *Domestics: UK domestic workers and their reluctant employers*, London: Work Foundation.

World Bank (various dates) Reports. Available <http://www-wds.worldbank.org/> (accessed 6 February 2005).

Worpole, K. and Greenhalgh, L. (1999) *The Richness of Cities*, London: Demos/Comedia.

Yeates, N. (2003) 'Social security in a global context', in J. Millar (ed.) *Understanding Social Security*, Bristol: Polity Press.

Yelling, J. (1986) *Slums and Slum Clearance in Victorian London*, London: Allen and Unwin.

Young, J. (1999) *The Exclusive Society: social exclusion, crime and social difference in modern society*, London: Sage.

Young, M. and Willmott, P. (1957) *Family and Kinship in East London*, London: Routledge.

Zuege, A. (1999) 'The Chimera of the Third Way', in L. Panitch and C. Leys (eds) *Socialist Register 2000*, London: Merlin.

INDEX

Page references in italic refer to illustrations.

eBooks – at www.eBookstore.tandf.co.uk

A library at your fingertips!

eBooks are electronic versions of printed books. You can store them on your PC/laptop or browse them online.

They have advantages for anyone needing rapid access to a wide variety of published, copyright information.

eBooks can help your research by enabling you to bookmark chapters, annotate text and use instant searches to find specific words or phrases. Several eBook files would fit on even a small laptop or PDA.

NEW: Save money by eSubscribing: cheap, online access to any eBook for as long as you need it.

Annual subscription packages

We now offer special low-cost bulk subscriptions to packages of eBooks in certain subject areas. These are available to libraries or to individuals.

For more information please contact webmaster.ebooks@tandf.co.uk

We're continually developing the eBook concept, so keep up to date by visiting the website.

www.eBookstore.tandf.co.uk